Algebra: The Basics of Mathematics

Algebra: The Basics of Mathematics

Joanna Burke

Larsen & Keller
www.larsen-keller.com

Algebra: The Basics of Mathematics
Joanna Burke
ISBN: 978-1-64172-492-0 (Hardback)

Larsen & Keller
Published by Larsen and Keller Education,
5 Penn Plaza,
19th Floor,
New York, NY 10001, USA

Cataloging-in-Publication Data

Algebra : the basics of mathematics / Joanna Burke.
p. cm.
Includes bibliographical references and index.
ISBN 978-1-64172-492-0
1. Algebra. 2. Mathematics. I. Burke, Joanna.
QA155 .A44 2020
512--dc23

Printed and bound in China.

For more information regarding Larsen and Keller Education and its products, please visit the publisher's website www.larsen-keller.com

Table of Contents

Preface

Algebra is one of the major areas of study in mathematics, along with geometry, number theory and analysis. Algebra studies mathematical symbols as well as the regulations which govern their manipulation. It is also considered as the unifying thread of almost the entirety of mathematics. In terms of complexity, algebra is divided into two parts, elementary algebra and abstract algebra. Elementary algebra, which is considered to be integral to the study of mathematics, deals with the basic aspects of algebra. Polynomials and polynomial expressions are examples of subjects which are studied under elementary algebra. The abstract parts of algebra fall under abstract algebra, which is also known as modern algebra. Sets, identity elements, inverse elements and associativity are some concepts studied within abstract algebra. The objective of this book is to give a general view of the different areas of algebra. It consists of contributions made by international experts. This book will serve as a reference to a broad spectrum of readers.

A foreword of all Chapters of the book is provided below:

Chapter 1 - The branch of mathematics which focuses on the study of mathematical symbols along with the rules for the manipulation of these symbols is referred to as algebra. Some of the major components of algebra are algebraic expressions and algebraic equations. This is an introductory chapter which will introduce briefly all the different operations in algebra such as addition, multiplication, subtraction and division; **Chapter 2** - Abstract algebra is a domain of algebra which deals with the study of algebraic structures. Some of the various algebraic structures are semigroups, monoids, groups, quasi-group and rings. The diverse applications of these algebraic structures in abstract algebra have been thoroughly discussed in this chapter; **Chapter 3** - The domain of algebra which deals with the study of linear equations and linear functions along with their representations through vector spaces and matrices is termed linear algebra. This chapter closely examines the key concepts of linear algebra such as determinant of a matrix, transpose of a matrix and inverse of a matrix to provide an extensive understanding of the subject; **Chapter 4** - The set of the basic concepts of algebra which are generally used for introducing variables is termed as elementary algebra. Some of the major concepts which are dealt with in elementary algebra are real numbers, exponents, polynomials and linear equations. This chapter discusses in detail these concepts related to elementary algebra; **Chapter 5** - There are a number of important theorems in algebra such as Cayley-Hamilton Theorem, Chebotarëv's theorem on roots of unity, the remainder theorem and the factor theorem, the fundamental theorem of ring homomorphisms and fundamental theorem of Galois Theory. The diverse applications of these theorems have been thoroughly discussed in this chapter.

I would like to thank the entire editorial team who made sincere efforts for this book and my family who supported me in my efforts of working on this book. I take this opportunity to thank all those who have been a guiding force throughout my life.

Joanna Burke

1

Introduction to Algebra

The branch of mathematics which focuses on the study of mathematical symbols along with the rules for the manipulation of these symbols is referred to as algebra. Some of the major components of algebra are algebraic expressions and algebraic equations. This is an introductory chapter which will introduce briefly all the different operations in algebra such as addition, multiplication, subtraction and division.

Algebra is a branch of mathematics that deals with properties of operations and the structures these operations are defined on. Elementary Algebra that follows the study of arithmetic is mostly preoccupied with operations on sets of whole and rational numbers and solving first and second order equations. What puts elementary algebra a step ahead of elementary arithmetic is a systematic use of letters to denote generic numbers.

Algebra is a branch of mathematics that deals in representing numbers through variables. Algebra also deals with symbols, relations, functions, and equations.

Algebraic Expressions

An algebraic expression is an expression formed from any combination of numbers and variables by using the operations of addition, subtraction, multiplication, division, exponentiation (rising to powers), or extraction of roots.

For instance,

$$7x,\ 2x-3y+1,\ \frac{5x^3-1}{4xy+1},\ \pi r^2\ and\ \pi r\sqrt{r^2+h^2}$$

are algebraic expressions. By an algebraic expression in certain variables we mean an expression that contains only those variables, and by a constant we mean an algebraic expression that contains no variables at all. If numbers are substituted for the variables in an algebraic expression, the resulting number is called the value of the expression for these values of the variables.

Example: Find the value of $\frac{2x-3y+1}{xy^2}$ when x = 2 and y = − 1.

Solution: Substituting x = 2 and y = − 1, we obtain,

$$\frac{2(2)-3(-1)+1}{2(-1)^2}=\frac{4+3+1}{2(1)}=\frac{8}{2}=4$$

If an algebraic expression consists of parts connected by plus or minus signs, it is called an algebraic sum, and each part, together with the sign preceding it, is called a term. For instance, in the algebraic sum,

$$3x^2y - \frac{4xz^2}{y} + \pi x^{-1}y,$$

the terms are $3x^2y$, $-4xz^2/y$, and $\pi x^{-1}y$.

Any part of a term that is multiplied by the remaining part is called a coefficient of the remaining part. For instance, in the term $-4xz^2/y$, the coefficient of z^2/y is $-4x$, whereas the coefficient of xz^2/y is -4. A coefficient such as -4, which involves no variables, is called a numerical coefficient. Terms such as $5x^2y$ and $-12x^2y$, which differ only in their numerical coefficients, are called like terms or similar terms.

An algebraic expression such as $4\pi r^2$ can be considered an algebraic sum consisting of just one term. Such a one-termed expression is called a monomial. An algebraic sum with two terms is called a binomial, and an algebraic sum with three terms is called a trinomial. For instance, the expression $3x^2 + 2xy$ is a binomial, whereas $-2xy^{-1} + 3\sqrt{x} - 4$ is a trinomial. An algebraic sum with two or more terms is called a multinomial.

A polynomial is an algebraic sum in which no variables appear in denominators or under radical signs, and all variables that do appear are raised only to positive-integer powers. For instance, the trinomial $-2xy^{-1} + 3\sqrt{x} - 4$ is not a polynomial; however, the trinomial $3x^2y^4 + \sqrt{2}xy - \frac{1}{2}$ is a polynomial in the variables x and y. A term such as $-1/2$, which contains no variables, is called a constant term of the polynomial. The numerical coefficients of the terms in a polynomial are called the coefficients of the polynomial. The coefficients of the polynomial above are $3, \sqrt{2},$ and $-1/2$.

The degree of a term in a polynomial is the sum of all the exponents of the variables in the term. When adding exponents, you should regard a variable with no exponent as being a first power. For instance, in the polynomial $9xy^7 - 12x^3yz^2 + 3x - 2$, the term $9xy^7$ has degree $1 + 7 = 8$, the term $-12x^3yz^2$ has degree $3 + 1 + 2 = 6$, and the term $3x$ has degree 1. The constant term, if it is zero, is always regarded as having degree 0.

The highest degree of all terms that appear with nonzero coefficients in a polynomial is called the degree of the polynomial. For instance, the polynomial considered above has degree 8. Although the constant monomial 0 is regarded as a polynomial, this particular polynomial is not assigned a degree.

Example: In each case, identify the algebraic expression as a monomial, binomial, trinomial, multinomial, and/or polynomial and specify the variables involved. For any polynomials, give the degree and coefficients.

a. $\frac{4x}{y}$

b. $4x + 3y^{-1}$

c. $-\frac{5}{3}x^2y + 8xy - 11$

Solution:

a. Monomial in x and y (not a polynomial because of the variable y in the denominator).

b. binomial in x and y, multinomial (not a polynomial because of the negative exponent on y).

c. trinomial, multinomial, polynomial in x and y of degree 3 with coefficients −5/3, 8, and −11.

A polynomial of degree n in a single variable x can be written in general form;

$$a_n x^n + a_{n-1} x^{n-1} + ... + a_2 x^2 + a_1 x + a_0$$

in which a_n, a_{n-1},..., a_2, a_1, a_0 are the numerical coefficients $a_n \neq 0$ (although any of the other coefficients can be zero), and a_0 is the constant term.

Algebraic Equation

In algebra, an equation can be defined as a mathematical statement consisting of an equal symbol between two algebraic expressions that have the same value.

The most basic and common algebraic equations in math consist of one or more variables.

For instance, 3x + 5 = 14 is an equation, in which 3x + 5 and 14 are two expressions separated by an 'equal' sign.

In an algebraic equation, the left-hand side is equal to the right-hand side.

Here, for example, 5x + 9 is the expression on the left-hand side, which is equal to the expression 24 on the right-hand side.

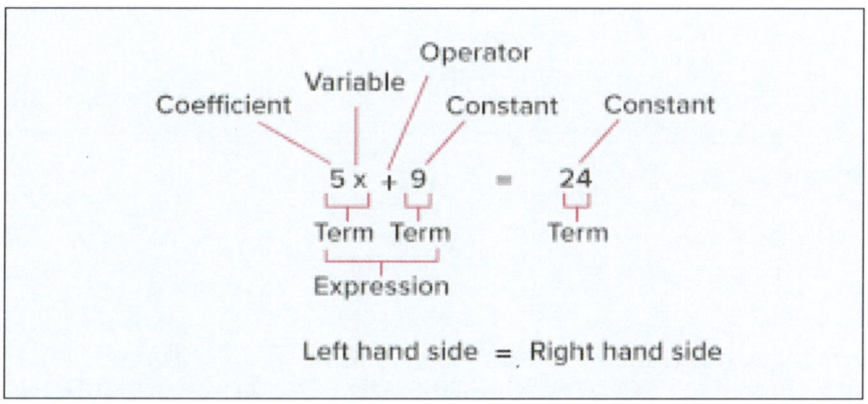

The process of finding the value of the variable is called solving the equation.

Like and Unlike Algebraic Terms

$9x + 4y^2 = 5$ is an algebraic equation. It consists of 3 terms i.e. $9x$, $4y^2$ and 5. The first two terms consists of variables and 5 is a constant.

$9x + 4y^2$ is an algebraic expression. It has two terms $9x$ and $4y^2$.

Therefore algebraic terms are those individual elements in an equation or an expression separated by '+' or '-' signs.

Consider the expression: $9x + 6y$

We cannot simplify this any further as 'x' and 'y' are unknown. Check out another example.

$$4\times2 + 3x + 4y + 8x + 10x^2$$

If rearranged, this expression can be re-written as,

$$10x^2 + 4x^2 + 8x + 3x + 4y$$

which is equal to,

$$14x^2 + 11x + 4y$$

Thus, it is seen that algebraic terms with same variables are added to each other. The addition of certain terms was possible only because the variables in both these cases are the same even if the numerical coefficients are different which can be added as normal numbers and the variable factor remains as it is. Now these terms which have the same variables are called like terms.

Thus, terms having identical variables raised to same exponent are like terms.

So, what are $14x^2$, $11x$ and $4y$ called? These are called unlike terms since the variables or exponent raised to these variables are "unlike" or not same.

Further operations on unlike terms cannot be directly performed. Based on this, it is clear that an algebraic expression consists of terms which can be categorized into like terms and unlike terms.

Let's consider another example $2xy + 4x^2 + 5xy + 5y^2 + 16x^2$

If you notice properly, the terms $2xy$ and $5xy$, as well as $4x^2$ and $16x^2$ have common factors. Only the numerical coefficients are different. Apart from that, all the variable factors are the same, so these terms can be added,

$$2xy + 5xy = 7xy$$

$$4x^2 + 16x^2 = 20x^2$$

Because these terms have the variable factors in common and an arithmetic operation can be performed on them they are called like terms. After adding the like terms, consider the expression again.

$$2xy + 4x^2 + 5xy + 5y^2 + 16x^2 = 7xy + 20x^2 + 5y^2$$

If you observe the expression now, none of the terms have any common factors and no arithmetic operation can be performed on them further. These terms are called unlike terms.

Like Terms	Unlike Terms
$2x + 19x$	$2x + 19a$
$4w - 10w$	$4w - 10w^2$
$14.2r - 12r$	$12r - 12s$
$32a^2 + 9a^2$	$32a^2 + 9a^3$
$8y + 5y$	$8y + 5$

Addition in Algebra

Addition of Like Terms

The rules of addition of like terms are:

Rule I: When all the terms are positive, add their coefficients, also the variables and power of the like terms remains the same.

Examples:

1. 5xy, 4xy and xy.

   Here, 5xy, 4xy and xy are like terms.

   The sum of the coefficients = 5 + 4 + 1 [xy means 1xy].

   Therefore, $5xy + 4xy + xy = 10xy$

   Note: To add two or more like terms, we add the numerical coefficients of the given terms and form another like term with the sum obtained as the numerical coefficient of the resulting term.

2. $5x + 4x + 2y + 3y$

   Here, 2x and 3x are like terms and also 6y and 5y are like terms.

   $5x + 4x = 9x$

   $2y + 3y = 5y$

   Therefore, the answer is $9x + 5y$

3. $3x^3 + 7x^3 + 4y^2 + 7y^2$

   Here, $3x^3 + 7x^3$ are like terms and also $4y^2 + 7y^2$ are like terms.

   $3x^3 + 7x^3 = 10x^3$

   $4y^2 + 7y^2 = 11y^2$

Therefore, the answer is $10x^3 + 11y^2$

Rule II: When all the terms are negative, add their coefficient without considering their negative signs and then prefix the minus sign (-) to the sum.

Examples:

1. -3ab, –5ab and –ab

   Without considering the negative signs, the coefficients of the given terms are 3, 5 and 1 respectively; and 3 + 5 + 1 = 9.

   Therefore, addition of -3ab, -5ab and –ab = –9ab

   i.e. (–3ab) + (–5ab) + (–ab) = –9ab

2. -5x + (–4x) + (–2y) + (–3y)

   Here, -5x and -4x are like terms and also -2y and -5y are like terms.

   (–5x) + (–4x) = –9x

   (–2y) + (–3y) = –5y

   Therefore, the answer is –9x - 5y.

Rule III: When all the terms are not of same sign. The same rule as that for the addition of integers should be applied.

Examples:

1. Addition of 21m and –9m

   = 21m + (-9m)

   = 21m - 9m

   = m(21 - 9)

   = 12m

2. 9xy – 4xy – 5xy + 7xy – xy

   = 5xy – 5xy + 7xy – xy

   = 0 + 7xy – xy, [since, 5xy – 5xy = 0]

   = 6xy.

Addition of Unlike Terms

Addition of Two Positive Unlike Terms

Suppose, to find the sum of two unlike terms x and y, we need to connect both the terms by using an addition symbol and express the result in the form of x + y.

Therefore, the sum of two unlike terms x and y = x + y.

Addition of Positive and Negative Unlike Terms

Suppose, to find the sum of two unlike terms x and -y, we need to connect both the terms by using an addition symbol [x + (−y)] and express the result in the form of x − y.

Therefore, the sum of two unlike terms x and -y = x + (−y) = x − y.

Addition of Negative and Positive Unlike Terms

Suppose, to find the sum of two unlike terms – x and y, we need to connect both the terms by using an addition symbol [(−x) + y] and express the result in the form of −x + y.

Therefore, the sum of two unlike terms -x and y = (−x) + y = −x + y.

Addition of Negative and Negative Unlike Terms

Suppose, to find the sum of two unlike terms −x and −y, we need to connect both the terms by using an addition symbol [(−x) + (−y)] and express the result in the form of −x − y.

Therefore, the sum of two unlike terms -x and -y = (−x) + (−y) = −x − y.

For example:

1. The unlike terms 2ab and 4bc cannot be added together to form a single term.

   All that which can be done is to connect them by the sign of addition and leave the result in the form 2ab + 4bc.

2. 5x + 3y + 2x + 3x.

   = 5x + 2x + 3x + 3y,

   = 10x + 3y, [Here 3y is an unlike term]

3. $3x^3 + 7y$

   Here $3x^3$ and 7y both are unlike terms so it will remain as it is.

   Therefore, the answer is $3x^3 + 7y$.

Multiplication in Algebra

Think of similar other situations in which two algebraic expressions have to be multiplied. Ameena gets up. She says, “We can think of area of a rectangle.” The area of a rectangle is l × b, where l is the length, and b is breadth. If the length of the rectangle is increased by 5 units, i.e., (l + 5) and breadth is decreased by 3 units, i.e., (b – 3) units, the area of the new rectangle will be (l + 5) × (b – 3).

Look at the following patterns of dots.

Pattern of dots	Total number of dots
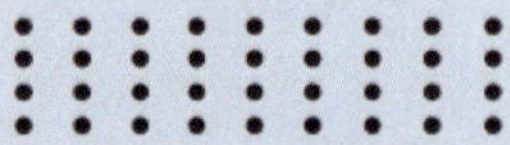	4×9
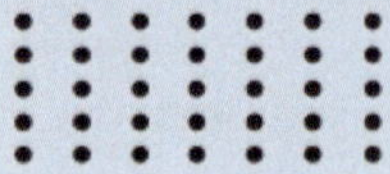	5×7
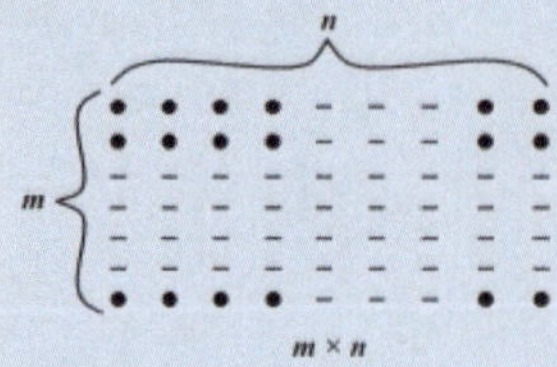	$m \times n$
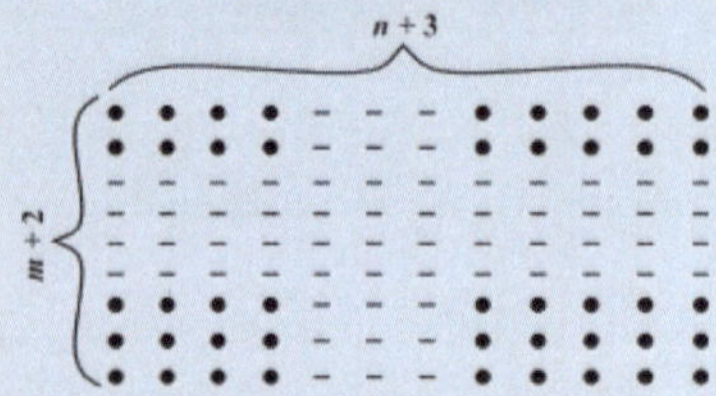	$(m + 2) \times (n + 3)$

Think about volume (The volume of a rectangular box is given by the product of its length, breadth and height).

Sarita points out that when we buy things, we have to carry out multiplication. For example, if:

- Price of bananas per dozen = Rs p;
- And, for the school picnic bananas needed = z dozens;
- Then, we have to pay = Rs p × z.

Suppose, the price per dozen was less by Rs 2 and the bananas needed were less by 4 dozens.

- Then, price of bananas per dozen = Rs (p – 2);
- And bananas needed = (z – 4) dozens;
- Therefore, we would have to pay = Rs (p – 2) × (z – 4).

In all the examples, we had to carry out multiplication of two or more quantities. If the quantities are given by algebraic expressions, we need to find their product. This means that we should know how to obtain this product. Let us do this systematically. To begin with we shall look at the multiplication of two monomials.

Multiplying a Monomial by a Monomial

Multiplying Two Monomials

We begin with $4 \times x = x + x + x + x = 4x$. Similarly, $4 \times (3x) = 3x + 3x + 3x + 3x = 12x$.

Now, observe the following products.

- $x \times 3y = x \times 3 \times y = 3 \times x \times y = 3xy$
- $5x \times 3y = 5 \times x \times 3 \times y = 5 \times 3 \times x \times y = 15xy$
- $5x \times (-3y) = 5 \times x \times (-3) \times y = 5 \times (-3) \times x \times y = -15xy$
- $5x \times 4x^2 = (5 \times 4) \times (x \times x^2) = 20 \times x^3 = 20x^3$
- $5x \times (-4xyz) = (5 \times -4) \times (x \times xyz) = -20 \times (x \times x \times yz) = -20x^2yz$

Observe how we collect the powers of different variables in the algebraic parts of the two monomials. While doing so, we use the rules of exponents and powers.

Multiplying Three or More Monomials

Observe the following examples:

- $2x \times 5y \times 7z = (2x \times 5y) \times 7z = 10xy \times 7z = 70xyz$
- $4xy \times 5x^2y^2 \times 6x^3y^3 = (4xy \times 5x^2y^2) \times 6x^3y^3 = 20x^3y^3 \times 6x^3y^3 = 120x^3y^3 \times x^3y^3$
  $= 120\,(x^3 \times x^3) \times (y^3 \times y^3) = 120x^6 \times y^6 = 120x^6y^6$

It is clear that we first multiply the first two monomials and then multiply the resulting monomial by the third monomial. This method can be extended to the product of any number of monomials.

Multiplying a Monomial by a Polynomial

Multiplying a Monomial by a Binomial

Let us multiply the monomial 3x by the binomial 5y + 2, i.e., find 3x × (5y + 2) = ? Recall that 3x and (5y + 2) represent numbers.

Therefore, using the distributive law,

$$3x \times (5y + 2) = (3x \times 5y) + (3x \times 2) = 15xy + 6x$$

We commonly use distributive law in our calculations. For example:

$7 \times 106 = 7 \times (100 + 6) = 7 \times 100 + 7 \times 6$ (Here, we used distributive law)

$= 700 + 42 = 742 \Rightarrow 7 \times 38 = 7 \times (40 - 2)$

$= 7 \times 40 - 7 \times 2$ (Here, we used distributive law)

$= 280 - 14 = 266$

Multiplying a Monomial by a Trinomial

Consider $3p \times (4p^2 + 5p + 7)$. We use distributive law;

$3p \times (4p^2 + 5p + 7) = (3p \times 4p^2) + (3p \times 5p) + (3p \times 7) = 12p^3 + 15p^2 + 21p$. Multiply each term of the trinomial by the monomial and add products.

By using the distributive law, we are able to carry out the multiplication term by term.

Multiplying a Polynomial by a Polynomial

Multiplying a Binomial by a Binomial

Let us multiply one binomial (2a + 3b) by another binomial, say (3a + 4b). We do this step-by-step, following the distributive law of multiplication,

$$(3a + 4b) \times (2a + 3b) = 3a \times (2a + 3b) + 4b \times (2a + 3b)$$

$$= (3a \times 2a) + (3a \times 3b) + (4b \times 2a) + (4b \times 3b)$$

$$= 6a^2 + 17ab + 12b^2 \text{(Since } ba = ab\text{)}$$

When we carry out term by term multiplication, we expect 2 × 2 = 4 terms to be present. But two of these are like terms, which are combined, and hence we get 3 terms. In multiplication of polynomials with polynomials, we should always look for like terms, if any, and combine them.

Multiplying a Binomial by a Trinomial

In this multiplication, we shall have to multiply each of the three terms in the trinomial by each of the two terms in the binomial. We shall get in all 3 × 2 = 6 terms, which may reduce to 5 or less, if the term by term multiplication results in like terms.

Consider,

$$\underbrace{(a+7)}_{\text{binomial}}\times\underbrace{(a^2+3a+5)}_{\text{trinomial}} = a\times(a^2+3a+5)+7\times(a^2+3a+5)\left[\text{using the distributive law}\right]$$

$$= a^3+3a^2+5a+7a^2+21a+35$$

$$= a^3+(3a^2+7a^2)+(5a+21a)+35$$

$$= a^3+10a^2+26a+35.$$

Subtraction in Algebra

Subtraction of Like Terms

In subtraction of like terms when all the terms are negative, subtract their coefficients, also the variables and power of the like terms remains the same.

For examples:

1. Subtract xy from 10xy.

   Here, 10xy, xy are like terms.

   The difference of the coefficients = 10 – 1, [xy means 1xy]

Therefore, 10xy - xy = 9xy

The result of subtraction of two like terms is also a like terms whose numerical coefficient is obtained by taking the difference of the numerical coefficients of like terms.

2. Subtract 4x from – 8x

   Here, 4x and -8x are like terms.

   = – 8x – (4x)

   = – 8x - 4x, [open the parenthesis]

   = – 12x

3. Subtract -3x from –7x

   Here, –3x and –7x are like terms.

   = –7x – (–3x)

   = –7x + 3x, [since negative times negative is positive so, –(–3x) = +3x]

   = –4x

4. 15x – 4x – 12y – 3y

   Here, 15x and 4x are like terms and also 12y and 3y are like terms.

   15x – 4x = 11x

   12y – 3y = 9y

   Therefore, the answer is 11x – 9y.

5. Subtract 4x + 3y + z from 2x + 3y – z.

   (2x + 3y – z) – (4x + 3y + z)

   = 2x + 3y – z – 4x – 3y – z, [open the parenthesis]

   Here, 4x and 2x are like terms,

   3y and –3y are like terms and also z and –z are like terms.

   Now arranging the like terms, we get,

   = 2x – 4x + 3y – 3y – z – z

   = –2x + 0 – 2z, [Since, + 3y – 3y = 0]

   = –2x – 2z

The subtraction of two or more like terms is another like term whose numerical coefficient is the subtraction of the numerical coefficients of these like terms.

Subtraction of Unlike Terms

The subtraction of unlike terms cannot be subtracted. For example, a – b will remain same as it is. Suppose the difference between two like terms is a single like term; but the two unlike terms cannot be subtracted to get a single term.

Subtraction of Two Positive Unlike Terms

To find the difference of two positive unlike terms suppose, take n from m, we need to connect both the terms by using a subtraction sign and express the result in the form of m – n.

Therefore, the difference of two positive unlike terms m and n = m – n.

Subtraction of Positive and Negative Unlike Terms

To find the difference of a positive and a negative unlike terms suppose, take -n from m, we need to connect both the terms by using a subtraction sign [m – (–n)] and express the result in the form of m + n.

Therefore, the difference of a positive and a negative unlike terms m and – n = m + n.

Subtraction of Negative and Positive Unlike Terms

To find the difference of a negative and a positive unlike terms suppose, take n from -m, we need to connect both the terms by using a subtraction sign [(–m) – n] and express the result in the form of –m – n.

Therefore, the difference of a negative and a positive unlike terms – m and n = – m – n.

Subtraction of Negative and Negative Unlike Terms

To find the difference of two negative unlike terms suppose, take -n from -m, we need to connect both the terms by using a subtraction sign [(–m) – (–n)] and express the result in the form of – m + n.

Therefore, the difference of two negative unlike terms -m and – n = – m + n.

For examples:

1. The unlike terms 2ab and 4bc cannot be subtracted to form a single term. All that which can be done is to connect them by the sign of subtraction and leave the result in the form 2ab – 4bc.

2. 11x – 7y –2x – 3x

   = 11x – 2x – 3x – 7y

   = 6x – 7y (here 7y is an unlike term).

3. 3x – 7y

   Here 3x and 7y both are unlike terms so it will remain as it is.

Therefore, the answer is 3x – 7y.

4. 15x – 12y – 11x

= 15x – 11x – 12y

= 4x – 12y (here 12y is an unlike term).

Division in Algebra

Our first examples of division of algebraic expressions involve simplifying and cancelling.

Examples:

1. Simplify $\frac{3ab\left(4a^2b^5\right)}{8a^2b^3}$

   First, we multiply out the top line:

   $$\frac{12a^3b^6}{8a^2b^3}$$

   When we write it out in full, this means:

   $$\frac{12\times aaa\times bbbbbb}{8\times aa\times bbb}$$

   Next, cancel the numbers top and bottom (we divide top and bottom by 4), the "*a*" terms (we cancel $a^2 = aa$ from top and bottom) and the "*b*" terms (we cancel $b^3 = bbb$ from top and bottom) to give us the final answer:

   $$\frac{3ab^3}{2}$$

2. Simplify $\frac{12m^2n^3}{\left(6m^4n^5\right)^2}$

   We square the denominator (bottom) of the fraction:

   $$\frac{12m^2n^3}{\left(6m^4n^5\right)^2}=\frac{12m^2n^3}{36m^8n^{10}}$$

   Next, we cancel out the numbers, and the "m" and "n" terms to give the final answer:

   $$\frac{1}{3m^6n^7}$$

3. Simplify $\frac{6p^3q^2-10p^2q}{4q}$

With this example, we'll break it into 2 fractions, both with denominator 4q to make it easier to see what to do.

$$\frac{6p^3q^2-10p^2q}{4q}=\frac{6p^3q^2}{4q}-\frac{10p^2q}{4q}$$

Next, we cancel the numbers and variables:

$$\frac{3p^3q}{2}-\frac{5p^2}{2}$$

Finally, we combine the fractions:

$$\frac{3p^3q-5p^2}{2}$$

After you have had some practice with these, you'll be able to do it without separating them into 2 fractions first.

Dividing by a Fraction

Recall the following when dividing algebraic expressions.

The reciprocal of a number x, is $\frac{1}{x}$.

For example, the reciprocal of 5 is $\frac{1}{5}$ and the reciprocal of $1\frac{2}{3}$ is $\frac{3}{5}$.

To divide by a fraction, you multiply by the reciprocal of the fraction.

For example,

$$\frac{3}{4}\div\frac{7}{x}=\frac{3}{4}\times\frac{x}{7}=\frac{3x}{28}$$

Examples:

Simplify $\dfrac{3+\frac{1}{x}}{\frac{5}{x}+4}$

Solution: Multiplying by the Reciprocal.

$$3+\frac{1}{x}=\frac{3x+1}{x}$$

We do likewise with the bottom expression (denominator):

$$\frac{5}{x}+\quad=\frac{5\quad 4}{x}$$

So the question has become:

$$\frac{3+\frac{1}{x}}{\frac{5}{x}+4}=\frac{\frac{3x+1}{x}}{\frac{5+4x}{x}}$$

We think of the right side as a division of the top by the bottom:

$$\frac{3x+1}{x}\div\frac{5+4x}{x}$$

To divide by a fraction, you multiply by the reciprocal:

$$\frac{3x+1}{x}\times\frac{x}{5+4x}=\frac{3x+1}{5+4x}$$

The x's cancelled out, and we have our final answer, which cannot be simplified any more.

Solution - Multiplying Top and Bottom.

$$\frac{3+\frac{1}{x}}{\frac{5}{x}+4}\times\frac{x}{x}$$

$$\frac{3+\frac{1}{x}}{\frac{5}{x}+4}\times\frac{x}{x}=\frac{3x+1}{5+4x}$$

It cannot be simplified further.

Long Division in Algebra

- Step 1: Make sure the polynomial is written in descending order. If any terms are missing, use a zero to fill in the missing term (this will help with the spacing).
- Step 2: Divide the term with the highest power inside the division symbol by the term with the highest power outside the division symbol.
- Step 3: Multiply (or distribute) the answer obtained in the previous step by the polynomial in front of the division symbol.
- Step 4: Subtract and bring down the next term.
- Step 5: Repeat Steps 2, 3, and 4 until there are no more terms to bring down.
- Step 6: Write the final answer. The term remaining after the last subtract step is the remainder and must be written as a fraction in the final answer.

Examples:

1. Divide: $\dfrac{x^3-4x^2+2x-3}{x+2}$

Step 1: Make sure the polynomial is written in descending order. If any terms are missing, use a zero to fill in the missing term (this will help with the spacing). In this case, the problem is ready as is.

$$x+2\overline{)x^3-4x^2+2x-3}$$

Step 2: Divide the term with the highest power inside the division symbol by the term with the highest power outside the division symbol. In this case, we have x^3 divided by x which is x^2.

$$\begin{array}{r} x^2 \\ x+2\overline{)x^3-4x^2+2x-3} \end{array}$$

Step 3: Multiply (or distribute) the answer obtained in the previous step by the polynomial in front of the division symbol. In this case, we need to multiply x^2 and $x + 2$.

$$\begin{array}{r} x^2 \\ x+2\overline{)x^3-4x^2+2x-3} \\ x^3+2x^2 \end{array}$$

Step 4: Subtract and bring down the next term.

$$\begin{array}{r} x^2 \\ x+2\overline{)x^3-4x^2+2x-3} \\ \underline{x^3+2x^2} \\ -6x^2+2x \end{array}$$

Step 5: Divide the term with the highest power inside the division symbol by the term with the highest power outside the division symbol. In this case, we have $-6x^2$ divided by x which is $-6x$.

$$\begin{array}{r} x^2-6x \\ x+2\overline{)x^3-4x^2+2x-3} \\ \underline{x^3+2x^2} \\ -6x^2+2x \end{array}$$

Step 6: Multiply (or distribute) the answer obtained in the previous step by the polynomial in front of the division symbol. In this case, we need to multiply $-6x$ and $x + 2$.

$$\begin{array}{r} x^2-6x \\ x+2\overline{)x^3-4x^2+2x-3} \\ \underline{x^3+2x^2} \\ -6x^2+2x \\ -6x^2-12x \end{array}$$

Step 7: Subtract and bring down the next term.

$$
\begin{array}{r}
x^2-6x \\
x+2\overline{)x^3-4x^2+2x-3} \\
\underline{x^3+2x^2} \\
-6x^2+2x \\
\underline{-6x^2-12x} \\
14-3
\end{array}
$$

Step 8: Divide the term with the highest power inside the division symbol by the term with the highest power outside the division symbol. In this case, we have 14x divided by x which is +14.

$$
\begin{array}{r}
x^2-6x+14 \\
x+2\overline{)x^3-4x^2+2x-3} \\
\underline{x^3+2x^2} \\
-6x^2+2x \\
\underline{-6x^2-12x} \\
14x-3
\end{array}
$$

Step 9: Multiply (or distribute) the answer obtained in the previous step by the polynomial in front of the division symbol. In this case, we need to multiply 14 and x + 2.

$$
\begin{array}{r}
x^2-6x+14 \\
x+2\overline{)x^3-4x^2+2x-3} \\
\underline{x^3+2x^2} \\
-6x^2+2x \\
\underline{-6x^2-12x} \\
14x-3 \\
14x+28
\end{array}
$$

Step 10: Subtract and notice there are no more terms to bring down.

$$
\begin{array}{r}
x^2-6x+14 \\
x+2\overline{)x^3-4x^2+2x-3} \\
\underline{x^3+2x^2} \\
-6x^2+2x \\
\underline{-6x^2-12x} \\
14x-3 \\
\underline{14x+28} \\
-31
\end{array}
$$

Step 11: Write the final answer. The term remaining after the last subtract step is the remainder and must be written as a fraction in the final answer.

$$x^2-6x+14-\frac{31}{x+2}$$

2. Divide: $\dfrac{-13x^2+4x^3+2x-7}{x^2+3x-2}$

Step 1: Make sure the polynomial is written in descending order. If any terms are missing, use a zero to fill in the missing term (this will help with the spacing). In this case, we should get:

$$x^2+3x-2\overline{)4x^3-13x^2+2x-7}$$

Step 2: Divide the term with the highest power inside the division symbol by the term with the highest power outside the division symbol. In this case, we have $4x^3$ divided by x^2 which is $4x$.

$$\begin{array}{r} 4x \\ x^2+3x-2\overline{)4x^3-13x^2+2x-7} \end{array}$$

Step 3: Multiply (or distribute) the answer obtained in the previous step by the polynomial in front of the division symbol. In this case, we need to multiply $4x$ and $x^2 + 3x - 2$.

$$\begin{array}{r} 4x \\ x^2+3x-2\overline{)4x^3-13x^2+2x-7} \\ 4x^3+12x^2-8x \end{array}$$

Step 4: Subtract and bring down the next term.

$$\begin{array}{r} 4x \\ x^2+3x-2\overline{)4x^3-13x^2+2x-7} \\ \underline{4x^3+12x^2-8x} \\ -25x^2+10x-7 \end{array}$$

Step 5: Divide the term with the highest power inside the division symbol by the term with the highest power outside the division symbol. In this case, we have $-25x^2$ divided by x^2 which is -25.

$$\begin{array}{r} 4x-25 \\ x^2+3x-2\overline{)4x^3-13x^2+2x-7} \\ \underline{4x^3+12x^2-8x} \\ -25x^2+10x-7 \end{array}$$

Step 6: Multiply (or distribute) the answer obtained in the previous step by the polynomial in front of the division symbol. In this case, we need to multiply -25 and $x^2 + 3x - 2$.

$$\begin{array}{r} 4x-25 \\ x^2+3x-2\overline{)4x^3-13x^2+2x-7} \\ \underline{4x^3+12x^2-8x} \\ -25x^2+10x-7 \\ -25x^2-75x+50 \end{array}$$

Step 7: Subtract and notice there are no more terms to bring down.

$$\begin{array}{r} 4x-25 \\ x^2+3x-2\overline{)4x^3-13x^2+2x-7} \\ \underline{4x^3+12x^2-8x} \\ -25x^2+10x-7 \\ \underline{-25x^2-75x+50} \\ 85x-57 \end{array}$$

Step 8: Write the final answer. The term remaining after the last subtract step is the remainder and must be written as a fraction in the final answer.

$$4x-25+\frac{85x-57}{x^2+3x-2}$$

3. Divide: $\dfrac{3x^3-2x^2+5}{x^2-1}$

Step 1: Make sure the polynomial is written in descending order. If any terms are missing, use a zero to fill in the missing term (this will help with the spacing). In this case, we should get:

$$x^2+0x-1\overline{)3x^3-2x^2+0x+5}$$

Step 2: Divide the term with the highest power inside the division symbol by the term with the highest power outside the division symbol. In this case, we have $3x^3$ divided by x^2 which is $3x$.

$$\begin{array}{r} 3x \\ x^2+0x-1\overline{)3x^3-2x^2+0x+6} \end{array}$$

Step 3: Multiply (or distribute) the answer obtained in the previous step by the polynomial in front of the division symbol. In this case, we need to multiply $3x$ and $x^2 - 1$.

$$\begin{array}{r} 3x \\ x^2+0x-1\overline{)3x^3-2x^2+0x+6} \\ 3x^3+0x^2-3x \end{array}$$

Step 4: Subtract and bring down the next term.

$$\begin{array}{r} 3x \\ x^2+0x-1\overline{)3x^3-2x^2+0x+6} \\ \underline{3x^3+0x^2-3x} \\ -2x^2+3x+6 \end{array}$$

Step 5: Divide the term with the highest power inside the division symbol by the term with the highest power outside the division symbol. In this case, we have $-2x^2$ divided by x^2 which is -2.

$$\begin{array}{r} 3x-2 \\ x^2+0x-1\overline{)3x^3-2x^2+0x+6} \\ \underline{3x^3+0x^2-3x} \\ -2x^2+3x+6 \end{array}$$

Step 6: Multiply (or distribute) the answer obtained in the previous step by the polynomial in front of the division symbol. In this case, we need to multiply -2 and $x^2 - 1$.

$$\begin{array}{r} 3x-2 \\ x^2+0x-1\overline{)3x^3-2x^2+0x+6} \\ \underline{3x^3+0x^2-3x} \\ -2x^2+3x+6 \\ -2x^2+0x+2 \end{array}$$

Step 7: Subtract and notice there are no more terms to bring down.

$$\begin{array}{r} 3x-2 \\ x^2+0x-1\overline{)3x^3-2x^2+0x+6} \\ \underline{3x^3+0x^2-3x} \\ -2x^2+3x+6 \\ \underline{-2x^2+0x+2} \\ 3x+4 \end{array}$$

Step 8: Write the final answer. The term remaining after the last subtract step is the remainder and must be written as a fraction in the final answer.

$$3x-2+\frac{3x+4}{x^2-1}$$

4. Divide: $\dfrac{2x^3+2x+7x^2+9}{2x+3}$

Step 1: Make sure the polynomial is written in descending order. If any terms are missing, use a zero to fill in the missing term (this will help with the spacing). In this case, we should get:

$$2x+3\overline{)2x^3+7x^2+2x+9}$$

Steps 2, 3, and 4: Divide the term with the highest power inside the division symbol by the term with the highest power outside the division symbol. Next multiply (or distribute) the answer obtained in the previous step by the polynomial in front of the division symbol. In this case, we should get $2x^3/2x = x^2$ and $x^2(2x + 3)$. Finally, subtract and bring down the next term.

$$\begin{array}{r|l} & x^2 \\ 2x+3 & 2x^3+7x^2+2x+9 \\ & \underline{2x^3+3x^2} \\ & \quad\quad\; 4x^2+2x \end{array}$$

Steps 5, 6, and 7: Divide the term with the highest power inside the division symbol by the term with the highest power outside the division symbol. Next multiply (or distribute) the answer obtained in the previous step by the polynomial in front of the division symbol. In this case, we should get $4x^2/2x = 2x$ and $2x(2x + 3)$. Finally, subtract and bring down the next term.

$$\begin{array}{r|l} & x^2+2x \\ 2x+3 & 2x^3+7x^2+2x+9 \\ & \underline{2x^3+3x^2} \\ & \quad\quad\; 4x^2+2x \\ & \quad\quad\; \underline{4x^2+6x} \\ & \quad\quad\quad\quad -4x+9 \end{array}$$

Steps 8, 9, and 10: Divide the term with the highest power inside the division symbol by the term with the highest power outside the division symbol. Next multiply (or distribute) the answer obtained in the previous step by the polynomial in front of the division symbol. In this case, we should get $-4x/2x = -2$ and $-2(2x + 3)$. Finally, subtract and notice there are no more terms to bring down.

$$\begin{array}{r|l} & x^2+2x-2 \\ 2x+3 & 2x^3+7x^2+2x+9 \\ & \underline{2x^3+3x^2} \\ & \quad\quad\; 4x^2+2x \\ & \quad\quad\; \underline{4x^2+6x} \\ & \quad\quad\quad\quad -4x+9 \\ & \quad\quad\quad\quad \underline{-4x-6} \\ & \quad\quad\quad\quad\quad\quad 15 \end{array}$$

Step 11: Write the final answer. The term remaining after the last subtract step is the remainder and must be written as a fraction in the final answer.

$$x^2+2x-2+\frac{15}{2x+3}$$

5. Divide: $\dfrac{x^3+3x^2-4x-12}{x^2+x-6}$

Step 1: Make sure the polynomial is written in descending order. If any terms are missing, use a zero to fill in the missing term (this will help with the spacing). In this case, the problem is ready as is.

$$x^2+x-6\overline{)x^3+3x^2-4x-12}$$

Steps 2, 3, and 4: Divide the term with the highest power inside the division symbol by the term with the highest power outside the division symbol. Next multiply (or distribute) the answer obtained in the previous step by the polynomial in front of the division symbol. In this case, we should get $x^3/x^2 = x$ and $x(x^2 + x - 6)$. Finally, subtract and bring down the next term.

$$\begin{array}{r} x \\ x^2+x-6\overline{)x^3+3x^2-4x-12} \\ \underline{x^3+x^2-6x} \\ 2x^2+2x-12 \end{array}$$

Steps 5, 6, and 7: Divide the term with the highest power inside the division symbol by the term with the highest power outside the division symbol. Next multiply (or distribute) the answer obtained in the previous step by the polynomial in front of the division symbol. In this case, we should get $2x^2/x^2 = 2$ and $2(x^2 + x - 6)$. Finally, subtract and notice there are no more terms to bring down.

$$\begin{array}{r} x+2 \\ x^2+x-6\overline{)x^3+3x^2-4x-12} \\ \underline{x^3+x^2-6x} \\ 2x^2+2x-12 \\ \underline{2x^2+2x-12} \\ 0 \end{array}$$

Step 8: Write the final answer. The term remaining after the last subtract step is the remainder and must be written as a fraction in the final answer. In this case, there is no remainder, so you do not need to write the fraction.

$$x+2$$

Solving Algebraic Equations

An equation states that two quantities are equal. This means that the left hand and right hand sides of the equals sign are equivalent, they balance. The equation may contain an unknown quantity that we wish to find. In the equation, $5x + 10 = 20$, the unknown quantity is x. This means that 5 multiplied by something that is then added to 10, will equal 20.

- To solve an equation means to find all values of the unknown quantity so that they can be substituted to make the left side equal to the right side and vice versa.

- Each such value is called a solution, or alternatively a root of the equation. In the example above, the solution is $x = 2$ because when 2 is substituted, both the left side and the right side equal 20 ∴ the sides balance. The value $x = 2$ is said to satisfy the equation.
- Sometimes we are required to rearrange or transpose the equation to solve it: essentially what we do to one side we do to the other.
- Hence, as above $5x + 10 = 20$, we first rearrange by subtracting 10 from the LHS and the RHS.
    - $5x + 10\ (-10) = 20(-10)$; $5x = 10$.
    - Then, we rearrange further to get the x on its own; so we divide both LHS and RHS by 5.
    - $\frac{5x}{5} = \frac{10}{5}$; now we have a solution: $x = 2$.
    - To check if this is correct we substitute x *for* 2; $(5 \times 2) + 10 = 20 \therefore 10 + 10 = 20$.

Four Principles to Apply when Solving an Equation

- Work towards the variable: Our aim is to get the variable by itself on one side of the equation. (So for the above we would aim to get the x by itself on one side.)

  $(x =)$

- Use the opposite mathematical operation: thus to remove a constant or coefficient, we do the opposite on both sides:

  Opposite of $\times$ is $\div$

  Opposite of $+$ is $-$

  Opposite of x^2 *is* $\sqrt{x}$

- Maintain balance: "What we do to one side, we must do to the other side of the equation."

- Check: Substitute the value back into the equation to see if the solution is correct.

One-step Equations

1. Addition Example:

$$x + (-5) = 8$$

Step 1: The constant, (−5), is the first target. So we need to do the opposite of plus (−5) which is to subtract (−5) from the LHS and the RHS:

$$x + (-5) - (-5) = 8 - (-5)$$

$$x = 8 + 5$$

$$\therefore x = 13$$

Check,

$$13 + (-5) = 8$$

$$13 - 5 = 8$$

2. Subtraction Example:

$$x - 6 = (-4)$$

Step 1: The constant, (−6) is the target. The opposite of subtract 6 is to add 6.

$$x - 6 + 6 = (-4) + 6$$

So,

$$x = (-4) + 6$$

$$\therefore x = 2$$

Check,

$$2 - 6 = (-4)$$

Two-step Equations

The following equations require two steps to single out the variable.

1. Addition Example:

$$2x + 6 = 14$$

In words: What number can we double then add six so we have a total of fourteen?

Step 1: The constant, 6, is our first target. If we subtract 6 from both sides, we create the following equation:

$$2x + 6 - 6 = 14 - 6 \text{ (The Opposite of } + 6 \text{ is } -6\text{)}$$

$$2x = 8 \ (+6 - 6 = 0)$$

Step 2: The only number left on the same side as the variable is the coefficient, 2. It is our second target. If we divide both sides by two, we create the following equation: (Between the 2 and the x is an invisible multiplication sign): $\frac{2x}{2}=\frac{8}{2}$ (The opposite of 2x is ÷ 2)

$\therefore x=4$

Check, if we substitute 4 into the equation we have:

$2 \times 4 + 6 = 14$

$8 + 6 = 14$

$14 = 14$ (We are correct)

2. Subtraction Example:

Solve for j: $3j - 5 = 16$

Step 1: $3j - 5 = 16$ (The first target is 5)

$3j - 5 + 5 = 16 + 5$ (Opposite of -5 is $+ 5$)

Thus,

$3j = 21$

Step 2: The second target is 3 and the opposite of × 3 is ÷ 3)

$$\frac{3j}{3}=\frac{21}{3}$$

$\therefore j=7$

Check,

$3 \times 7 - 5 = 16$

3. Multi-step Example:

Solve for T : $\frac{3T}{12}-7=6$

$$\frac{3T}{12}-7=6 \text{ (Target 7 then 12 then 3)}$$

$$\frac{3T}{12}-7+7=6+7 \text{ (add 7 to LHS and RHS)}$$

$$\frac{3T}{12}=13 \text{ (now target 12 to get 3 T on its own)}$$

$$\frac{3T}{12}\times 12=13\times 12 \text{ (multiply LHS and RHS by 12)}$$

$$3T = 156\left(\text{now target 3 to get the T on it own}\right)$$

$$\frac{3T}{3} = \frac{156}{3}\left(\text{divide LHS and RHS by 3}\right)$$

$$\therefore T = 52$$

Check, $\left(3 \times 52\right) \div 12 - 7 = 6$

Solving Equations with Fractions

It is important to remember that in algebra, terms are separated by a plus (+) or minus (–) sign or by an equals (=) sign. Whereas variables that are multiplied or divided are considered one term.

For example, there are four terms in this equation: 8ab + 3ab + 4 = 70

And also, 3(x + 2) + 8 – 6 = 20

When working with fractions, it is generally easier to eliminate the fractions first.

Examples:

1. Let's solve for: $\frac{2}{3}x - 6 = 4$ (There are 3 terms in this equation?)

Step 1: Eliminate the fraction, to do this we work with the denominator first, rather than multiply by the reciprocal as this can get messy. So both sides of the equation will be multiplied by 5, which include distributing 5 through the brackets:

$$5\left(\frac{2}{5}x - 6\right) = 5 \times 4$$

$$2x - 30 = 20$$

Step 2: Target the constant: $2x - 30 + 30 = 20 + 30$ (The opposite of subtraction is addition)

$$2x = 50$$

Step 3: Target the variable: $\frac{2x}{2} = \frac{50}{2} \therefore x = 25$

Check,

$$\left(\frac{2}{5} \times \frac{25}{1}\right) - 6 = 4$$

$$(2 \times 5) - 6 = 4 \therefore 10 - 6 = 4$$

2. Let's solve for $x: \frac{3}{5}(x + 5) = 9$

Step 1: Eliminate the fraction: Again we work with the denominator and multiply everything by 5:

$$5\left[\frac{3}{5}(x + 5)\right] = 5(9)$$

The 5 is not distributed through the brackets $(x + 5)$ because $\frac{3}{5}(x+5)$ is one term.

The result,

$$3(x + 5) = 45$$

Step 2: Target the variable by dividing by 3: $\frac{3(x+5)}{3}=\frac{45}{3};x+5=15$

Now we target the $5: x+5-5=15-5 \therefore x=10$

Check,

$$\frac{3}{5}(10+5)=9;\frac{3}{5}\times\frac{15}{9}=9$$

Solving Equations with Variables on Both Sides

Solving equations with variables on both sides can be difficult and requires some methodical mathematical thinking. Remembering that both sides are equivalent, the goal is to get all of the constants on one side of the equation and the variables on the other side of the equation.

Examples:

1. Solve for: 3x + 5 + 2x = 12 + 4x

Work towards rearranging the equation to get all of the constants on the RHS and the variables on the LHS:

$$3x+5+2x-4x=12+4x-4x$$

Let's start by subtracting $4x$ from both sides.

The result,

$$3x+5+2x-4x=12$$

$$3x+5-5+2x-4x=12-5$$

Now subtract 5 from both sides.

$$3x+2x-4x=7$$

Now simplify by collecting like terms. (3 + 2 − 4 = 1)

$$1x=7 \therefore x=7$$

Check,

$$3x+5+2x=12+4x$$

21 + 5 + 14 = 12 + 28 (Substitute the variable for 7)

40 = 40

2. Solve for $x: 6x+3(x+2)=6(x+3)$

Again, work towards moving all the constants on one side and the variables on the other. First we could distribute through the brackets which will enable us to collect like terms:

$6x+3x+6=6x+18$ (Distribute through the brackets)

$6x+3x+6-6=6x+18-6$ (Subtract 6 from both sides first, to get constants on RHS)

The result,

$$6x+3x=6x+18-6$$

$9x-6x=6x-6x+18-6$ (Subtract 6*xx* from both sides to get variables on LHS)

$$3x=12 \therefore x=4$$

Check,

$$6x+3(x+2)=6(x+3); 24+18=6\times 7 \therefore 42=42$$

Multiples and Factors

Multiples, Common Multiples and the LCM

- A multiple of a given a number, into which that number can be divided exactly. For example: multiples of 5 are 5, 10, 15, 20, 25, 30...
- A common multiple is a multiple in which two or more numbers have in common. For example: 3 and 5 have multiples in common 15, 30, 45...
- The lowest common multiple LCM is the lowest multiple that two numbers have in common For example: 15 is the LCM of 3 and 5.

Factors, Common Factors and the HCF

- Factor: A whole number that can be multiplied a certain number of times to reach a given number.
    - 3 is factor of 15 because it can be multiplied by 5 to get 15.
- A common factor is a factor that two or more numbers have in common.
    - 3 is a common factor of 12 and 15.
- The highest common factor – HCF
    - What is the HCF of 20 and 18?
    - We list all of the factors first.
    - The factors of 20: (1, 2, 4, 5, 10, 20).

    - The factors of 18: (1, 2, 3, 6, 9, 18).
    - The common factors are 1 and 2 and the highest common factor is 2.
- Proper factors: All the factors apart from the number itself.
    - The factors of 18 (1, 2, 3, 6, 9, 18).
    - Thus the proper factors of 18 are 1, 2, 3, 6, 9.
- Prime number: Any whole number greater than zero that has exactly two factors – itself and one.
    - 2, 3, 5, 7, 11, 13, 17, 19...
- Composite number: Any whole number that has more than two factors.
    - 4, 6, 8, 9, 10, 12, 14, 15, 16, 18, 20...
- Prime factor: A factor that is also a prime number.
- Factor tree: A tree that shows the prime factors of a number As the factor trees show, some numbers have many factors. This is a helpful way to find the factors of a given number. A more effective method for clarity when thinking and reasoning mathematically is to list all of the factors.

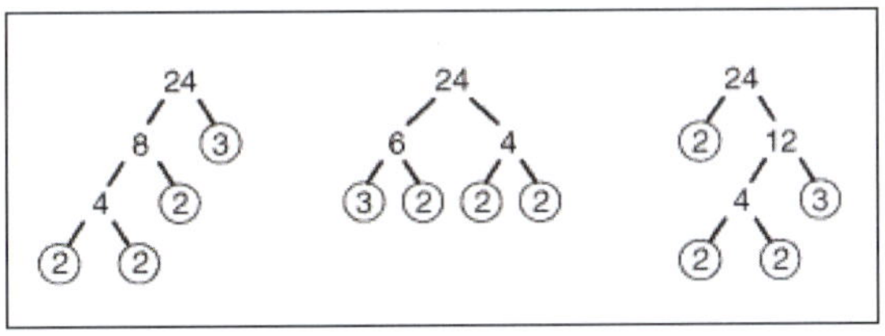

Factorising and Expanding

The distributive law is commonly used in algebra. It is often referred to as either 'expanding the brackets' or 'removing the brackets'.

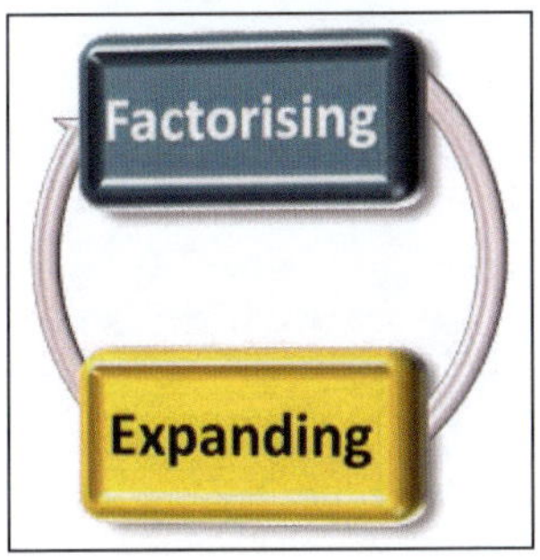

- Expanding involves taking what is outside of the brackets and moving it through the brackets. For example:

  $3(x + 2)$ can be expanded to $3 \times x + 3 \times 2$ which is $3x + 6$

This means that, $3(x + 2) = 3x + 6$

- Working in the opposite direction is called factorisation and this process is slightly more difficult.
- Factorisation requires finding the highest common factor. (The HCF then sits outside of the brackets.)
- Let's look at $3x + 6$, we can factorise this back again from the expanded form.
    - First we need to find a common 'factor'; we can see that x can be divided by 3 and 6 can be divided by 3. Our common factor is 3. Now we put this 3 outside of the brackets (because we have divided each of the terms by three) to be multiplied by everything inside of the brackets.
    - So we have $3(x + 2)$.
- Another factorisation example:
    - $5x + 15x^2 - 30x^3$ has three terms $5x$ and $15x^2$ and $30x^3$.
    - All three terms contain x and can be divided by 5 $\therefore$ $5x$ is a common factor.
    - Next we divide each term by $5x$.
    - $5x \div 5x = 1$; $15x^2 \div 5x = 3x$; and $30x^3 \div 5x = 6x^2$.
    - We end up with: $5(1 + 3x - 6x^2)$.
- Now we can do the reverse and practise expanding the brackets.
    - So expand $5x(1 + 3x - 6x^2)$.
    - Multiply each term inside the brackets by 5x.

$$5x \times 1 = 5x; 5x \times 3x = 5 \times 3 \times x \times x = 15x^2; \text{and } 5x \times 6x^2 = 5 \times 6 \times x \times x \times x = 30x^3.$$

    - Now we have expanded the brackets to get $5x + 15x^2 - 30x^3$.

Example:

Remove the brackets:

- $5x(2 + y)$ expands to $(5x \times 2) + (5x \times y)$ which is simplified to $10x + 5xy$.
- $-x(2x + 6)$ expands to $(-x \times 2\,x) + (-x \times 6)$ which is simplified to $-2x^2 - 6x$.

Factorise:

- $3x + 9x - x^2$ has only the variable x in common $\therefore$ $x(3 + 9 - x)$.
- $12x^3 + 4x^2 - 20x^4$ has $4x^2$ as the highest common factor $\therefore$ $4x^2(3x + 1 - 5x^2)$.

References

- What is Algebra, What is: cut-the-knot.org, Retrieved 1 August , 2019
- Introduction-to-algebraic-expression-and-polynomials, algebra, tutorials: emathzone.com, Retrieved 9 May, 2019
- Equation, number-sense, math-vocabulary: splashmath.com, Retrieved 8 August , 2019
- Like-and-unlike-algebraic-terms, maths: byjus.com, Retrieved 31 March, 2019
- Addition-of-Like-Terms: math-only-math.com, Retrieved 14 July, 2019
- Subtraction-of-Like-Terms: math-only-math.com, Retrieved 17 May, 2019
- 3-division-algebra, basic-algebra: intmath.com, Retrieved 19 April, 2019
- Long-division, divide-poly, notes, mat120, scotz47781: mesacc.edu, Retrieved 5 February, 2019

2

Abstract Algebra

Abstract algebra is a domain of algebra which deals with the study of algebraic structures. Some of the various algebraic structures are semi-groups, monoids, groups, quasi-group and rings. The diverse applications of these algebraic structures in abstract algebra have been thoroughly discussed in this chapter.

Abstract algebra is a broad field of mathematics, concerned with algebraic structures such as groups, rings, vector spaces, and algebras.

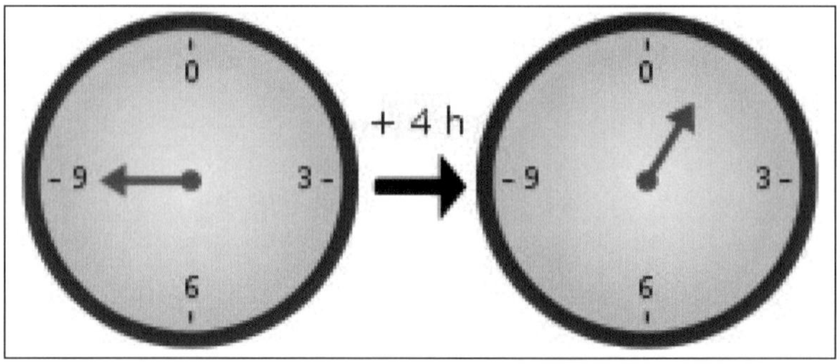

On the 12-hour clock, 9 + 4 = 19 + 4 = 1, rather than 13 as in usual arithmetic.

Roughly speaking, abstract algebra is the study of what happens when certain properties of number systems are abstracted out; for instance, altering the definitions of the basic arithmetic operations result in a structure known as a ring, so long as the operations are consistent.

For example, the 12-hour clock is an example of such an object, where the arithmetic operations are redefined to use modular arithmetic (with modulus 12). An even further level of abstraction-where only one operation is considered-allows the clock to be understood as a group. In either case, the abstraction is useful because many properties can be understood without needing to consider the specific structure at hand, which is especially important when considering the relationship(s) between structures; the concept of a group isomorphism is an example.

Levels of Abstraction in Abstract Algebra

It is possible to abstract away practically all of the properties found in the "usual" number systems, the tradeoff being that the resulting object-known as a magma (which consists of a set and a binary operation, that need not satisfy any properties other than closure)-is simply too general to be interesting. On the other extreme, it is possible to abstract out practically no properties, which allows for many results to be found, but the resulting object (the usual number systems) is too specific to solve more general problems.

Most of abstract algebra is dedicated to objects that have a reasonable balance between generality and structure, most notably groups and rings in which most of the basic *properties* of arithmetic are maintained, but their specifics are left free. Still, some higher levels of abstraction are occasionally useful; quasi-groups, for instance, are related to Latin squares, and monoids are often used in computer science and are simple examples of categories.

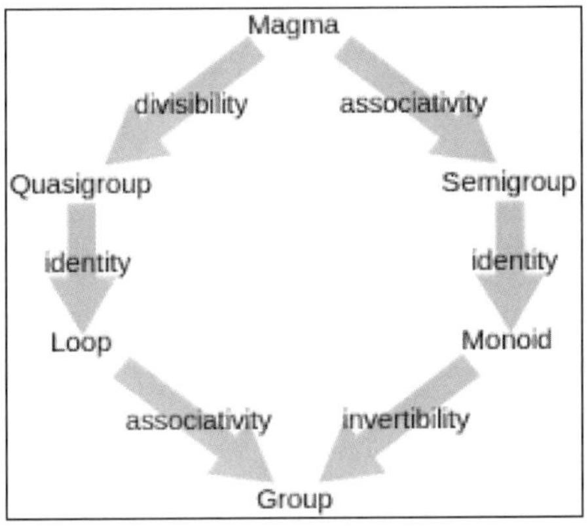

Algebraic Structure

A non-empty set G equipped with one or more binary operations is said to be an algebraic structure. Suppose * is a binary operation on G. Then (G, *) is an algebraic structure. (N,*), (1, +), (1, -) are all the algebraic structure. Here, (R, +, .) is an algebraic structure equipped with two operations.

Binary Operation on a Set

Suppose G is a non-empty set. The G X G = {(a, b): a E G, b E G)}. If f: G X G → G then f is called a binary operation on a set G. The image of the ordered pair (a, b) under the function f is denoted by afb.

A binary operation on asset G is sometimes also said to be the binary composition in the set G. If * is a binary composition in G then, a * b E G, a, b E G. Therefore g is closed with respect to the composition denoted by *.

Example:

An addition is a binary operation on the set N of natural number. The sum of two natural numbers is also a natural number. Therefore, N is a natural number with respect to addition i.e. a + b.

Subtraction is not a binary operation on N. We have 4 – 7 = 3 not belong to N whereas 4 E N. thus, N is not closed with respect to subtraction, but subtraction is a binary operation on the set of an integer.

Properties of an Algebraic Structure

By a property of an algebraic structure, we mean a property possessed by any of its operations. Important properties of an algebraic system are:

1. Associative and commutative laws:

   An operation * on a set is said to be associative or to satisfy the associative law if, for any elements a, b, c in S we have (a * b) * c = a * (b * c)

An operation * on a set S is said to be commutative or satisfy the commutative law if, a * b = b * a for any element a, b in S.

2. Identity element and inverse:

   Consider an operation * on a set S. An element e in S is called an identity elements for * if for any elements a in S - a * e = e * a = a.

   Generally, an element e is called a left identity or a right identity according to as e *a or a * e = a where a is any elements in S.

   Suppose an operation * on a set S does have an identity element e. The inverse of an element in S is an element b such that: a * b = b * a = e

3. Cancellation Laws

   An operation * on a set S is a said to satisfy the left cancellation law if, a * b = a * c implies b = c and is said to satisfy the right cancellation law if, b * a = c * a implies b = c.

Semi-groups

A set with one binary operation satisfying the law of associativity. A semi-group is a generalization of the concept of a group: only one of the group axioms is retained — associativity; this is the explanation of the term "semi-group". Semi-groups are called *monoids* if they have, in addition, an identity element.

There are an extraordinary number of examples of semi-groups. Among these are various sets of numbers which are closed under addition or multiplication; semi-groups of matrices with respect to multiplication; semi-groups of functions with respect to "point wise" multiplication $*$, defined by $(f * g)(x) = f(x)g(x)$; semi-groups of sets with respect to intersection or union; etc. The following example is important in the general theory and in some applications. Let X be an arbitrary set, and let an operation on the set FX of all finite sequences of elements of X be defined by the formula:

$$(x_1,\dots,x_n) * (y_1,\dots,y_m) = (x_1,\dots,x_n, y_1,\dots,y_m)$$

Then F_X is a semi-group relative to $*$; it is known as the free semi-group on X. Every semi-group is a homomorphic image of some free semi-group.

Any set of mappings of an arbitrary set M into itself, which is closed under composition (i.e. the operation of successive application, also known as superposition), is a semi-group with respect to that operation; this is the case, in particular, for the set of all self-mappings of a set M, known as the symmetric semi-group on M. Many important sets of transformations turn out to be semi-groups, and they often fail to be groups. On the other hand, any semi-group is isomorphic to some semi-group of mappings. Thus, the concept of a semi-group proves to be most suitable for the investigation of mappings in a quite general context, and it is largely through the consideration of mappings that the links between the theory of semi-groups and other fields of mathematics have been realized. In this framework, semi-groups appear very frequently as semi-groups of endomorphisms of the system being studied: spaces, algebras, graphs, etc. Semi-groups also appear in the theory of partial transformations and binary relations with respect to multiplication.

As in other algebraic theories, one of the main problems of the theory of semi-groups is the classification of all semi-groups and a description of their structure. This is achieved by imposing various restrictions on the semi-groups under consideration and thereby specifying various types of semi-groups. Such restrictions may be of different kinds. A semi-group may satisfy a fixed system of identities (typical examples are commutative semi-groups and semi-groups of idempotents) or other conditions, expressed by a formula of first-order predicate calculus (examples are semi-groups with a cancellation law and regular semi-groups). The cancellation law and regularity are examples of restrictions which in a sense constitute weak versions of group properties; the introduction of such conditions was particularly popular in the early days of the theory of semi-groups (among the most "group-like" types thus defined are right groups). In many cases, however, the classes of semi-groups obtained in this way include semi-groups with properties not at all similar to those of groups (typical examples are semi-groups of idempotents).

The concept of a regular semi-group arose in analogy with that of a regular ring (in the sense of von Neumann). Regular semi-groups are among the most intensively investigated in the theory of semi-groups. They include the following important classes: the multiplicative semi-groups of regular rings (in particular, the semi-group of all matrices of a given order over a division ring), symmetric semi-groups, the semi-groups of all partial transformations of sets, inverse semi-groups, Clifford semi-groups, in particular, semi-groups of idempotents and completely-simple semi-groups, completely 0-simple semi-groups, etc.

Another type of commonly imposed restriction are restrictions on the system of all or some of the sub-semi-groups, in particular the ideals, and also on certain relations on semi-groups, in particular congruences. Such restrictions give rise, for example, to various types of simple semi-groups and various finiteness conditions, semi-groups with different types of ideal series and ideal systems; an essential role in the investigation of many problems of the theory of semi-groups is played by the Green equivalence relations.

The restrictions may relate to generating sets, delineating various types according to the nature of the generating elements (e.g. idempotents; any semi-group can be imbedded in an idempotently-generated semi-group), the number of such elements (finitely-generated semi-groups play an essential part in many investigations), or the interaction of the generators — semi-groups given by defining relations and, in particular, finitely-presented semi-groups; finally, the types of generating sets may be specified in both of the above respects.

When the structure of semi-groups is considered, much importance is attached to various constructions that reduce the description of the semi-groups in question to that of "better" types. Quite frequently, the latter are groups, and the principle of description "modulo groups" is common in semi-group-theoretical contexts; in fact, it already appeared in the above-mentioned classical theorem of Rees, according to which any completely 0-simple (completely-simple) semi-group is isomorphic to a regular Rees matrix semi-group over a group with a zero (over a group). Groups take part in the constructions that describe inverse semi-groups, and in those that describe commutative Archimedean semi-groups with a cancellation law and without idempotents. The description of semi-groups with several finiteness conditions reduces to that of groups with the corresponding conditions.

Among the constructions figuring in the description of semi-groups one has both general algebraic ones, such as direct and subdirect products, and specific semi-group-theoretical ones. The latter include, besides the above-mentioned Rees semi-groups, various other constructions, such as that of a band — a partition into sub-semi-groups such that the corresponding equivalence relation is congruence. Particularly important bands are the commutative bands (or semi-lattices) and matrix (rectangular) bands. Many types of semi-groups can be described in terms of bands. Thus, Clifford's theorem for completely-regular semi-groups means, essentially, that these semi-groups are semi-lattices of completely-simple semi-groups; the completely-simple semi-groups are precisely the rectangular bands of groups; the Tamura–Kimura theorem states that any commutative semi-group admits a unique decomposition as a band of Archimedean semi-groups.

As always in algebra, the concept of a homomorphism plays an essential role also in the theory of semi-groups, and hence so does the concept of congruence. Semi-groups belong to the class of universal algebras whose congruence are not uniquely determined by any canonical coset ("kernel"), as is the case, say, for groups and rings. This more complicated situation has led to the development of a fairly extensive branch of the theory of semi-groups, devoted to the investigation of semi-group congruences from various points of view. The problems involved fall mainly into two categories: 1) to investigate some special type of congruence on arbitrary semi-groups; 2) to describe all congruences on some special semi-groups, belonging to some class of importance. The first category includes, in particular, the study of principal congruences, and also of ideal, or Rees, congruences, associated with the two-sided ideals of a semi-group (if I is an ideal in a semi-group S, then the corresponding Rees congruence classes are I itself and the singletons $\{x\}$, $x \in S\setminus I$). Rees congruences are frequently used in various problems and explain the importance of studying ideals; the quotient semi-group modulo Rees congruence is known as the Rees quotient semi-group modulo the corresponding ideal. Worthy of mentioning among the solved problems in the second category are the description of the congruences on symmetric semi-groups and on completely 0-simple semi-groups, and the far-reaching investigation of congruences on inverse semi-groups; the theory of radicals of semi-groups has been developed, not without the influence of the analogous branch of ring theory.

In connection with the theory of sub-semi-groups, one finds a self-contained branch of the theory dealing with the study of lattice properties of semi-groups, i.e. the relationship between the properties of semi-groups and the properties of their semi-group lattice.

Another extensive branch of the theory of semi-groups is connected with various imbedding of semi-groups. The roots of this branch go back to the classical problem of imbedding of semi-groups in groups. For some problems and results in this branch of the theory.

Intensive attention has been devoted to varieties of semi-groups; concerning this field of semi-groups. First steps have been taken towards a theory of quasi-varieties of semi-groups and certain other classes of semi-groups similar in some sense to varieties.

The general theory of semi-groups links up with specific semi-groups in many ways. Among the problems solved are the abstract characterization of various important concrete semi-groups (such as transformation semi-groups; in particular, there are several characterizations of symmetric semi-groups), and several of their abstract properties have been described. Consideration has been given to isomorphisms and homomorphisms of abstract semi-groups into various specific semi-groups, first and foremost — transformation semi-groups and matrix semi-groups. The investigation of homomorphisms of semi-groups into certain semi-groups of numbers, mainly into the multiplicative semi-group of complex numbers, forms the subject of the theory of semi-group characters.

One of the specific fields in the theory of semi-groups is the study of semi-groups with an additional structure compatible with the multiplication operation. In this context one should mention, above all, the structure of a topological space and the structure of a partial or total order.

Theories have also been developed for certain types of generalized semi-groups. These are primarily algebras with one n-ary operation satisfying a generalized associative law (known as n-associative or n-semi-group operations). Another variant is that of algebras with one partial associative binary operation (a natural situation of this type arises in the theory of categories).

Monoid

A monoid consists of a universe M together with an identity element $1 \in M$ in M and a multiplication operation, denoted multiplicatively, which is associative in the sense that:

$$(m \cdot n) \cdot k = m \cdot (n \cdot k)$$

holds for every m, n, k $\in$ M.

The order of evaluation in a monoid is not important, i.e. that different ways of parenthesising a sequence of elements in the monoid will yield the same result as far as monoid multiplication is concerned. This is made explicit in the following fact, which is implicitly presented to students already in primary school.

Proof: A parenthesisation can be viewed as a binary tree, where the leaves are labelled by $m_1, \ldots\ldots\ldots m_k$ when read from left to right. The associativity equation in the above definition says that doing a rotation does not affect the value of a tree. Every two trees that have the same sequence of leaves can be transformed into each other using a finite sequence of rotations.

Monoid Homomorphisms

A monoid homomorphism if a function between monoids that preserves the structure of monoids, i.e. a monoid homomorphism between monoids M and N is a function:

$$h : M \to N$$

which preserves the identity element and which is consistent with the multiplication operation in the sense that:

$$h(m \cdot n) = h(m) \cdot h(n)$$

An alternative definition of a homomorphism would be that the following diagram commutes,

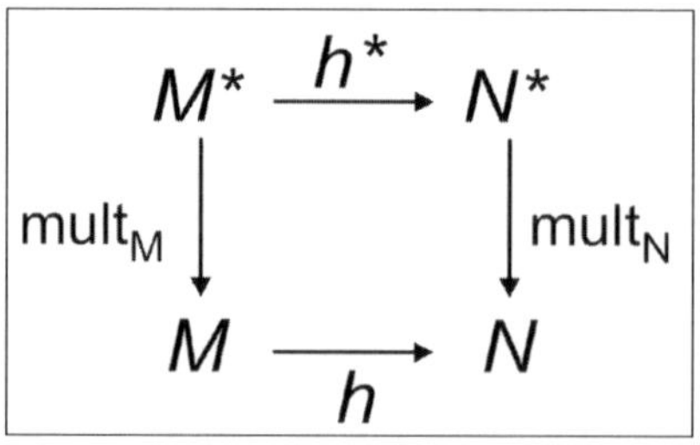

with the two downward arrows representing the multiplication operations in the respective monoids.

Using monoid homomorphisms, we can express the following universal property of the monoid Σ^* It is generated by Σ, and it is the biggest monoid generated by Σ in the following sense. For every monoid N that is generated by Σ, there exists a surjective monoid homomorphism,

$$h : \Sigma^* \rightarrow M$$

which is the identity on the Σ generators. This is property goes under the name Σ^* is the free monoid with generators Σ.

Monoid homomorphisms are closely related with functions that are compositional. Let M be a monoid, and let X be a set. A function:

$$h : M \rightarrow X$$

is called compositional if for every m, n $\in$ M, the value h(m $\cdot$ n) is uniquely determined by the values h(m) and h(n). Again, compositionality can be stated in terms of terms of a commuting diagram: h is compositional if there is some function $mult_X$ such that the following diagram commutes.

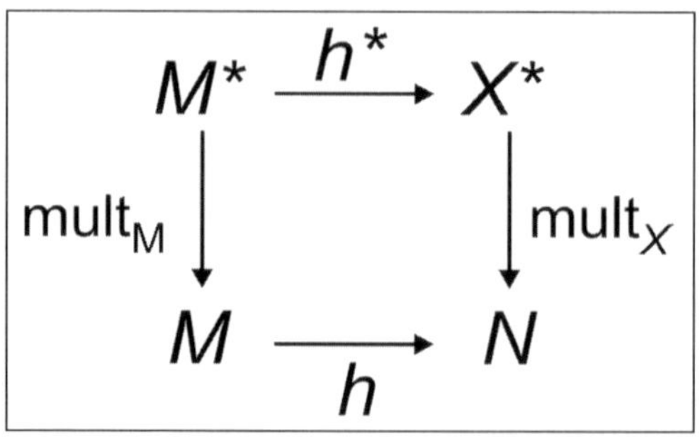

Lemma

Let M be a monoid, and let h : M $\rightarrow$ X be a sujrective compositional function. Then there exists (a unique) monoid structure on X which makes h into a monoid homomorphism.

Proof: Saying that h(m $\cdot$ n) is uniquely determined by h(m) and h(n), as in the definition of compositionally, means that there is a function:

$$f : X \times X \rightarrow X$$

such that every m, n $\in$ M satisfy,

$$h(m \cdot n) = f\left(h(m)h(n)\right).$$

One uses f to be the product operation in X, and takes the identity to be the image under h of the identity.

Recognising Languages

We will be interested in monoid as an alternative to finite automata, i.e. we will use monoids to recognise languages.

A language $L \subseteq A^*$ is said to be recognised by a monoid homomorphism:

$$h : A^* \rightarrow M$$

if membership in $w \in L$ is determined uniquely by h(w). In other words, there exists a subset $F \in M$ such that:

$$w \in L \quad \textit{if} \quad h(w) \in F$$

holds for every word $w \in A^*$.

The following theorem shows that, as far as recognising languages is concerned, finite monoids and finite automata are equivalent.

Theorem: The following conditions are equivalent for every language $L \subseteq A^*$:

- L is recognised by a finite automaton;
- L is recognised by a monoid homomorphism into a finite monoid.

Proof: From a monoid one creates a deterministic automaton, whose states are elements of the monoid, and where the transition relation is obtained from the monoid operation in the obvious way. This proves the bottom-up implication. For the top-down implication, let us transform a nondeterministic automaton into a monoid. Let A be a nondeterministic automaton with input alphabet A and states Q. Define the function,

$$\delta : A^* \rightarrow P(Q \times Q)$$

which sends a word w to the set of pairs (p, q) such that there exists a run of the automaton which begins in p, reads the word w, and ends in q. One can show that δ is compositional, and therefore there is a monoid structure on its image which makes it into a monoid homomorphism. The appropriate multiplication operation is simply composition of relations, while the identity is the identity relation.

Note how the transformation from a nondeterministic finite automaton to a monoid incurs an exponential blow-up. Note also that the transformation from nondeterministic finite automaton to a deterministic finite automaton is exponential, but going directly from a nondeterministic finite automaton to monoids is not doubly exponential.

The Syntactic Monoid

Deterministic finite automatons have minimisation, i.e. there is always a minimal deterministic automaton that can be obtained from any other automaton by quotienting. The same holds for monoids, as proved in the following theorem.

Theorem: For every language $L \subseteq A^*$ there exists a monoid homomorphism:

$$h : A^* \to M,$$

called the syntactic homomorphism of L, which recognises it and is minimal in the sense that for every surjective monoid homomorphism:

$$g : A^* \to N$$

which also recognises L, there exists a unique monoid homomorphism which makes the following diagram commute.

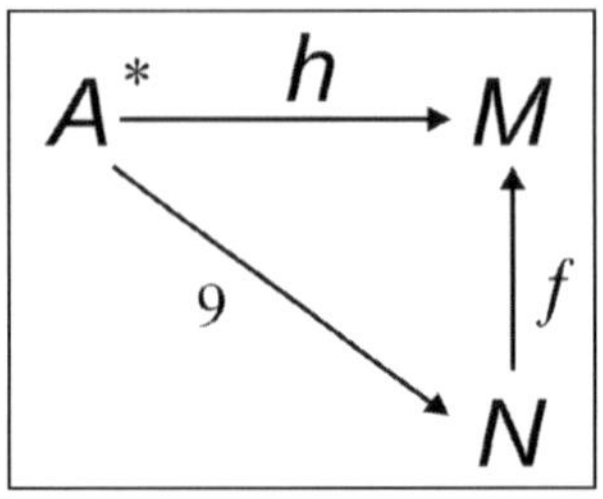

Proof: Define the syntactic equivalence of L to be the equivalence relation on A* which identifies two words $w, w' \in A^*$ if,

$$wuv \in L \quad if \quad wu'v \in L$$

holds for every $u, v \in A^*$. One can show that the function, call it h, which maps a word to its equivalence class, is compositional, and therefore by Lemma 1, the set of equivalence classes is equipped with a monoid structure that makes h into a monoid homomorphism. It is not difficult to check that h is minimal as required in the theorem.

Idempotent Power

An idempotent in a monoid is an element e such that ee = e. Typically, idempotent are denoted by e, f. The identity is an idempotent. An idempotent power of an element m of a monoid is an element of the form m_i that is idempotent. It is easy to see that the idempotent power, if it exists, is unique (in the sense that the monoid element is unique, not the exponent), because if m^i and m^j would be idempotent powers, then we would need to have:

$$m^i = m^{ij} = m^j$$

The following basic lemma shows that every element of a finite monoid has an idempotent power. This idempotent power will be denoted by $m^\#$ here.

1. Idempotent Power Lemma: If m is an element of a finite monoid of size n, then $m^{n!}$ is an idempotent.

Green's Relations

For a monoid M, define the following relations. We say that $m \in M$ is a prefix of $n \in M$ if there exists some $x \in M$ such that $n = mx$. This is the same thing as saying that the right ideal of m, i.e. the set mM, contains the right ideal nM. Likewise we define the suffix and infix relations, using the left ideal Mm and the two-sided ideal MmM, respectively. It is easy to check that all three relations are quasi-orderings, in the sense that they are transitive and reflexive, but not necessarily anti-symmetric.

Clearly a prefix is a special case of an infix. Therefore, every R-class, i.e. every equivalence class with respect to prefixes, is entirely included in some J-class, i.e. some equivalence class with respect to infixes. A key property of finite monoids is that inside a single J-class, the R-classes form an antichain.

2. Eggbox Lemma: Consider elements m, n of a finite monoid. If m is a prefix of n and they are both in the same J-class, then also n is a prefix of m.

   Proof: By the assumptions of the lemma, there exist x, y, z in the monoid such that $n = mx$ and $m = ynz$. By iterating these equalities, we see that:

   $$m = y^i m (xz)^i$$

   holds for every natural number i. If we choose i to be the idempotent power of the monoid, which exists by the assumption on finiteness, we see that:

   $$m = y^i m (xz)^i = y^i m (xz)^i (xz)^i = m (xz)^i = n (zx)^{i-1} z$$

   thus proving that n is a prefix of m.

   The above lemma, although simple, will be used very frequently in the following topic, and one could say that it essentially contains the theory of finite monoids.

   Call elements H-equivalent if they have the same R-class and the same L-class. Therefore an H-class is an intersection of an R-class and an L-class.

3. H-Dichotomy Lemma: Let H be an H-class included in a J-class J. Then either,

   - $mn \notin J$ for every m, $m \in H$; or
   - H is a group.

   Proof: Suppose that there exist $m_0, n_0 \in H$ such that $m_0 n_0 \in J$. We will prove that H is a group. (Note also that the alternative in the statement of the lemma is exclusive.)

   First let us show that $mn \in J$ holds for every $m, n \in H$. Since m is in the same L-class as m_0, there is some x such that $m_0 = xm$. Likewise, since n is in the same R-class as n_0, there is some y such that $n_0 = ny$. Therefore $mn \in J$ since it is an infix of some element of J, namely $mn = xm_0 n_0 y$.

Now let us show that H is a monoid, i.e. $mn \in H$ holds for every $m, n \in H$. Since m is a prefix of mn, then by the Eggbox Lemma, m is R-equivalent to mn. Likewise, n is L-equivalent to mn, and therefore $mn \in H$.

Finally let us show that H is a group.

By the Idempotent Power Lemma, H contains some idempotent, call it e. We will show that this idempotent is the identity with respect to multiplication in H, and is therefore unique. Let m \ in H. Because m and e are in the same L-class, there is some x such that m = xe. Therefore me = xee = xe = m. By a symmetric argument, em = m holds for every $m \in H$, and therefore e is the identity in H.

Let us now show that H has inverses. By the Idempotent Power Lemma, every $m \in H$ has some idempotent power, call it m^k. there can only be one idempotent in H. Therefore m^{k-1} is the inverse of m.

This completes the proof of the lemma.

A corollary of the H-dichotomy lemma is that an H-class is a group if and only if it contains an idempotent.

4. Lemma: The J-class of the identity is a group.

   Proof: Since the identity is an idempotent, the H-Dichotomy Lemma implies that the H-class of the identity is a group. We will now prove that the J-class of the identity is the same thing as the H-class of the identity. If m is in the J-class of the identity, then since the identity is a prefix of m, the Eggbox Lemma can be used to conclude that the identity is R-equivalent to m. Likewise for L-classes. Therefore, the J-class of the identity is actually an H-class.

First-order Logic

Consider the property "for every position that has label a, there is a later position that has label b". This property can be formalized in the language of logic as:

$$\forall x a(x) \Rightarrow \exists y (x < y \wedge b(y))$$

This is the idea behind first-order definable languages of words. One uses formulas, where the variables quantify over positions of the word, with a binary predicate $x < y$ that denotes the order on positions, and with unary predicates of the form a(x) to test the labels of positions. A set of words is called first-order definable if there is a formula of this kind that is true on words from the set and false in other words. The term "first-order" refer to the fact that the logic can only quantify over single positions, as opposed to quantification over sets of positions, or sets of sets of positions, which would be possible in higher-order logics (e.g. monadic second-order logic is allowed to quantify over sets of positions, but not sets of pairs of positions).

Example: The language (a + b)*b is definable in first-order logic, using the formula given above.

Star-free Expressions

Star-free expressions are, essentially, a minor variation on first-order logic. A star-free expression is a regular expression that does not use the star, but it can use complementation (with respect to an alphabet that is implicit in the expression). For example, if the input alphabet is Σ, then the complement of the empty set is a star-free expression that describes Σ^*. Using this expression, one can e.g. define the set of words that contain at least one letter a, this is:

$$(\neg\emptyset)\cdot a\cdot(\neg\emptyset)$$

Aperiodic Monoids

Call a finite monoid aperiodic if for every m in the monoid, the sequence:

$$m, m^2, m^3, \ldots.$$

is ultimately constant. It is easy to see that this is equivalent to saying that every m in the monoid satisfies:

$$m^{\#} = m^{\#}m$$

where $m^{\#}$ is the idempotent power.

Let S be a semigroup. Define an S-factorisation tree to be a tree where nodes are labelled by elements of S, subject to the following restrictions:

- If a node has children with labels $s_1, \ldots.., s_n$, read from left to right, then its label is the product $s_1, \ldots, s_n$;
- If a node has more than two children, all of these children have the same label, and this label is an idempotent.

Define the yield of a tree to be the sequence of leaf labels read from left to right (this is a word in S^+). Define the height of a tree to be the biggest number of edges on a root-to-leaf path. We are interested in creating S-factorisation trees of small height. Clearly every word of length n is the yield of some S-factorisation tree of height at most log n, which does not use idempotent nodes, and is constructed by recursively splitting the word in two parts. As the following example shows, using idempotent nodes can decrease the height of a tree to a constant, e.g. at most 6 for the morphism which counts a‘s modulo two.

Example: Consider the semigroup {0, 1} with addition modulo 2.

- Every word with only zeros has a factorisation tree of height at most 1, by combining all the zero leaves using an idempotent root.
- Every word with exactly two ones has a factorisation tree of height at most 4, by using binary nodes to combine the trees for the blocks of zeros with the two ones.
- Every word with an even number of ones has a factorisation tree of height at most 5, by combining the trees from the previous item using an idempotent root.
- Every word with an odd number of ones has a factorisation tree of height at most 6, by using the previous item and using a binary root to append the extra one with its trailing zeros.

Therefore, every word has a factorisation tree of height at most 6.

Theorem (Factorisation Tree Theorem)

For every finite semigroup S, there is a constant k such that every word in S^+ is the yield of an S-factorisation tree of height at most k.

For a semigroup S, define height(S) $\in \mathbb{N} \cup \{\infty\}$ to be the smallest number k such that every word in S^+ is the yield of some S-factorisation tree of height at most k. The theorem says that height(S) is finite for every finite S.

Consider a surjective semi-group morphism h: $S \to T$. If e is an idempotent in T, then its inverse image under h is easily seen to be closed under multiplication, and therefore is a sub semi-group of S.

1. Lemma: If h : $S \to T$ is a surjective semi-group morphism, then:

$$height(S) \leq height(T) \cdot \max_{e} height\left(h^{-1}(e)\right)$$

where e ranges over idempotents in T.

Proof: Let $w \in S^+$ be a word. First create a factorisation tree t for the word in T^+ obtained from w by applying h to every letter. Let us look at this tree as a candidate for an S-factorisation tree. The leaf nodes are correct, and the binary nodes are correct, because they do not depend on the structure of the semi-group. The only issue is with the idempotent nodes. Consider an idempotent node in the tree t, corresponding to some idempotent $e \in T. Let\ \omega_1, \ldots.. w_n$ be the infixes of w that correspond to the children of this idempotent node; their products in S belong to the sub semi-group $h^{-1}(e)$. Therefore we can combine them into an $h^{-1}(e)$-factorisation tree, which is also an S-factorisation tree.

Let us now prove the Factorisation Tree Theorem. The proof is by induction on the size of the semi-group. Let S be a semigroup. In order to use the induction assumption, we will try to use Lemma 1. Consider a morphism h like in Lemma 1, and the inequality in the conclusion of the lemma. If h is not a bijection, then the semigroup T is strictly smaller than S. If h has an image of size at least two, then every semigroup of the form $h^{-1}(e)$ is also smaller than S. Therefore, if h is nontrivial, i.e. it is not a bijection and has image of size at least two, then we can apply the induction assumption to all semigroups that appear on the right side of the inequality in Lemma 1.

Let J be a $\mathcal{J}$-class which is as simple as possible (unlike for monoids, a semigroup need not have a unique $\mathcal{J}$-class that is simplest). Let $J/\mathcal{H}$ be $\mathcal{H}$the set of H-classes in J and consider the function:

$$h : S \to \mathcal{J} / \mathcal{H}\{0\}$$

which maps each element of J to its $\mathcal{H}$-class, and which maps other elements to a designated zero element.

2. Lemma: The function h is compositional.

Proof: We need to show that h(s) and h(t) uniquely determine h(st). First observe that if either $h(s)$ or $h(t)$ is zero, which means that either s or t is outside J, then also $st \notin J$, by choice of $\mathcal{J}$ as lightest possible $\mathcal{J}$-class, and therefore $h(st)$ is zero. Consider now the case when $s, t \in J$. From Green's relations it follows that membership st $\in$ J is uniquely determined by the $\mathcal{H}$-classes of s and t, and so is the $\mathcal{H}$-class of st if $st \in J$.

From the above lemma it follows that h is a semigroup morphism, with an appropriate semigroup structure on its image. Therefore, if h is nontrivial then we can use the induction assumption and Lemma. The remaining case is when h is trivial, i.e. h maps all elements to a single one, or h is a bijection. When h maps all elements to a single one, this means that all of the semigroup is in a single $\mathcal{H}$-class, which means that the semigroup is a group, by the $\mathcal{H}$-dichotomy lemma.

The remaining case is when h is a bijection. This means that every $\mathcal{H}$-class in J is a singleton, and there is at most one element outside J, call this element zero. The element zero o, if it exists, satisfies so = os = o.

First observe, that the interesting case is words which have type other than zero. If a word has zero type, then it admits a decompositional,

$$w = w_1 s_1 w_2 s_2 \ldots w_n s_n w_{n+1}$$

where $w_1, \ldots . w_n \in S^+$ and $w_{n+1} \in S^*$ have nonzero type, and each s_i is chosen so that $w_i s_i$ has zero type. Therefore, a tree for w can be obtained by combining the zeros of $w_i s_i$ using an idempotent node, and then adding w_{n+1} using a binary node (if it is nonempty).

We are therefore left with finding trees of small height for words w $\in S^+$ that have nonzero type. We prove this by induction on the number of two-letter infixes, i.e. the following parameter:

$$[w] = \{st \in S^2 : \text{w has an infix st}\}$$

Note that these infixes cannot use the zero, so they are actually in J^2.

The base case is when w has no two-letter infixes, i.e. it has length one, and therefore a factorisation tree of height zero. For the induction step, choose some two letter infix $st \in J^2$. Let,

$$w = w_0 st w_1\ st w_2 \ldots w_n\ st w_{n+1}$$

be a decomposition of w which identifies all appearances of the infix *st*. By induction assumption, trees of small height can be found for each w_i used in the above decomposition. If $n \in \{0,1\}$, then these trees can be combined using the binary rule. Suppose that $n \geq 2$, and consider the words:

$$tw_1\ s, \ldots tw_n s.$$

They all have the same $\mathcal{R}$-class, namely that of t, and they all have the same $\mathcal{L}$-class, namely that of s, and therefore they all have the same type by assumption on all $\mathcal{H}$-classes being singletons. Furthermore, this type is idempotent, because also $tw_1 stw_2\ s$ has this type.

A monoid consists of a universe M together with a product operation:

$$\pi : M^* \to M$$

which maps each one letter word m to m, and which is associative in the sense that the following diagram commutes,

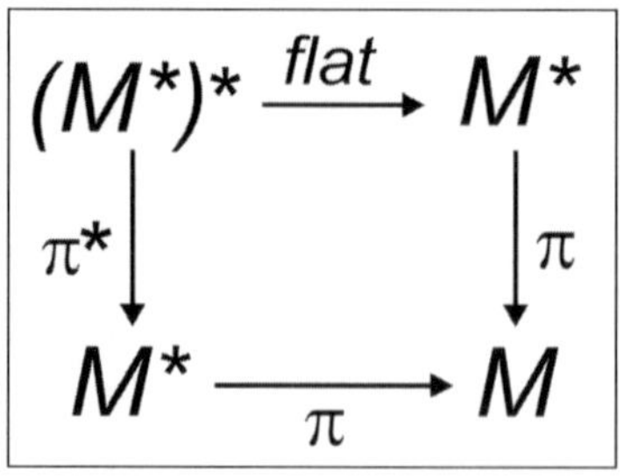

where flat flattens a word of words into a word in the obvious way, while $\pi*$ applies the product operation to every word in a word of words.

To define monoids for infinite words, we use the same definition, only * is replaced by ∞. This requires a new flattening operation:

$$flat : \left(M^{\infty}\right)^{\infty} \rightarrow M^{\infty}$$

which we illustrate on three examples:

$$\begin{array}{ll} (aba)(\in)(aa)(b) & \mapsto\; abaaab \\ (aaa....)(bbb...) & \mapsto\; aaa... \\ (\in)(\in)(\in) & \mapsto\; \in \end{array}$$

Note how in the second example above, the first infinite word causes the rest of the input to be ignored. This ignoring is a little bit of a hack, which could be avoided by either using two sorts (finite and infinite word), or by considering a richer notion of words which is closed under substitution, e.g. words where the positions form a countable well-ordering.

An ∞-monoid consists of a unierse M together with a product operation:

$$\pi : M^{\infty} \rightarrow M$$

which is the identity on one letter words and which is associative in the sense that the following diagram commutes.

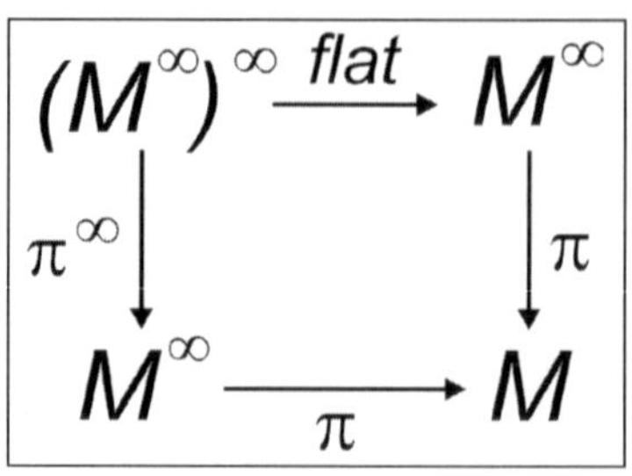

It is not difficult to see that $\sum^{\infty}$, with flattening as the monoid operation, is an ∞-monoid. Here is an example of an ∞-monoid with a finite universe.

Example: This ∞-monoid is going to recognize the ∞-words over alphabet {a, b} which contain infinitely many a' s. The universe is:

$$M = \{\in, a, b, a^{\omega}, b^{\omega}\}$$

The product operation is defined as follows. To define $\pi(w)$ with $w \in M^{\infty}$, we do the following. First, remove all ε letters from w. If the result is empty, return ε. If there is a letter of the form a^{ω} or b^{ω}, then return the a^{ω} or b^{ω}, whichever comes first in w. Otherwise, if the result is an infinite word, then return a^{ω} if there are infinitely many a's and b^{ω} otherwise. Otherwise, return a if there is at least one a and b otherwise.

Homomorphism

Define a homomorphism of ∞-monoids M and N to be a function h between their universes which respects the structure of ∞-morphisms in the sense that the following diagram commutes.

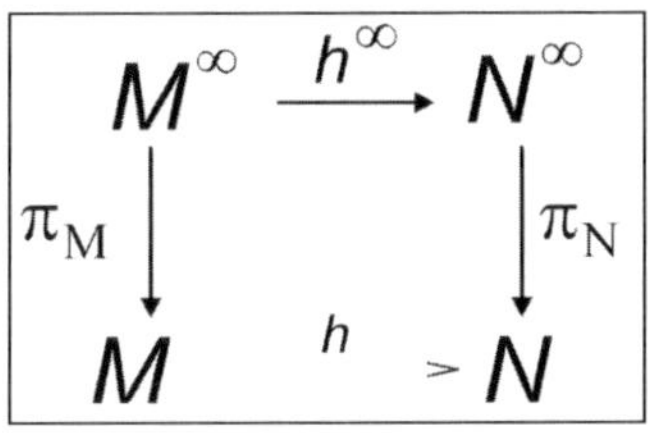

Compositional Functions

As with normal monoids, instead of defining an associative product operation on a universe N, it is sometimes easier to give a compositional function from an existing ∞-monoid to N, and get the ∞-monoid structure on N automatically from that. We describe this now. Suppose that M is an ∞ -monoid and N is a set. A function:

$$h : M \rightarrow N$$

is called compositional if every words $w, v \in M^{\infty}$ satisfy

$$h^{\infty}(\omega) = h^{\infty}(v) \text{ implies } h(\pi(w)) = h(\pi(v))$$

Lemma; If h: M $\rightarrow$ N is compositional, then there is a multiplication operation on N which turns it into an ∞-monoid, and which turns h into a homomorphism of ∞-monoids.

Proof: Compositionality can be rephrased as saying that there is some function σ which makes the following diagram commute.

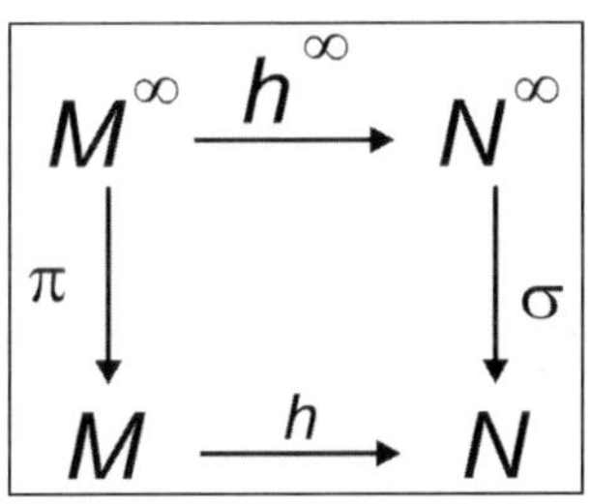

The diagram then shows that h is a homomorphism, assuming that σ yields a structure of an ∞-monoid on N, i.e. it is associative.

A monad consists of the following ingredients:

- For each set X, a new set T X, called the structures over X.
- For each function f : X \to Y, a new function $T\,f : T\,X \rightarrow T\,Y$, called the lifting of f.
- For each set X, a flattening operation $\text{flatt}_X : TT\,X \rightarrow T\,X$.
- For each set X, a unit operation $\text{unit}_X : X \rightarrow T\,X$.

The above definition is actually a special case of the general definition of monads. Monads can be applied to any category, while the above definition is only given for the category of sets.

Example: The monad of nonempty finite words is defined as follows. The structures over X are X^+, i.e. the nonempty finite words over X. The lifting f^+ (w) is obtained by applying f to each letter of w. The flattening operation is the natural one which takes a word of words into a single word, e.g. the flattening of (aba) (aa) (b) is abaaab. The unit operation maps a letter to a single letter word consisting of this letter.

Functor Axioms

The first group of axioms is that the first two ingredients of a monad are a functor in the sense of category theory. This means lifting is preserves the identity and composition of functions. Preserving identity means that if we lift the identity function on X, the result is the identity function on T X. Preserving composition means that for every functions f: $X \rightarrow Y$ and g: $Y \rightarrow Z$, the following diagram commutes.

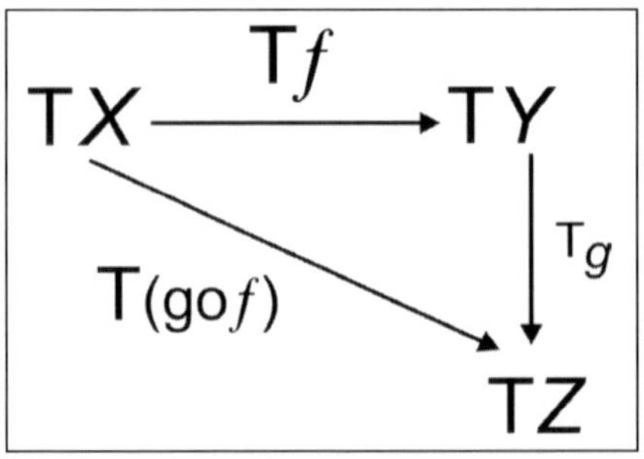

Natural Axioms

The second group of axioms is that for every function f $X \rightarrow Y$, the following diagrams commute. This means that flattening and unit are what category theorists call natural transformations.

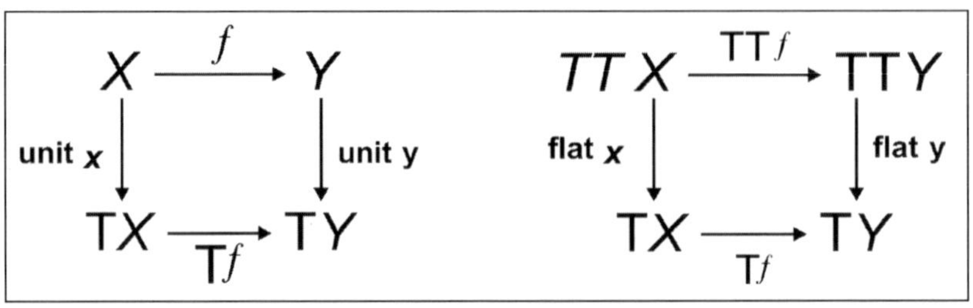

Monoid Axioms

The first monoid axiom says, essentially, that applying the unit to a structure (i.e. an element of T X) and flattening has no effect. The expression "applying the unit to a structure" is ambiguous in the sense that it has two different interpretations, and both interpretations are dealt with in the axiom, which says that the following diagram commutes for every set X.

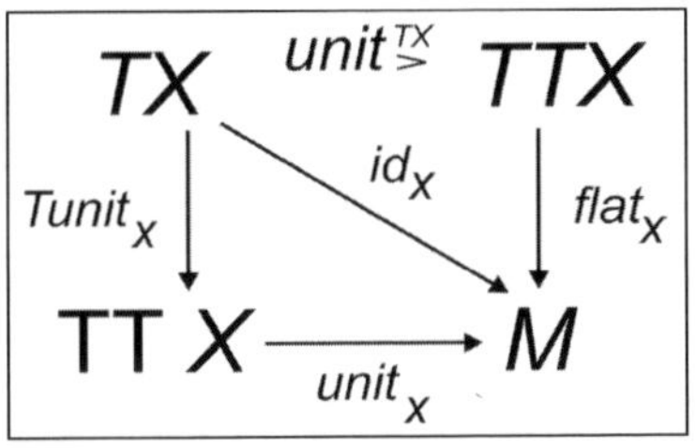

The second monoid axiom is the most important one. It says that if we start with a structure of structures of structures, and we flatten it twice, then we get the same result regardless of which order of flattening we employ. The appropriate commuting diagram is given below.

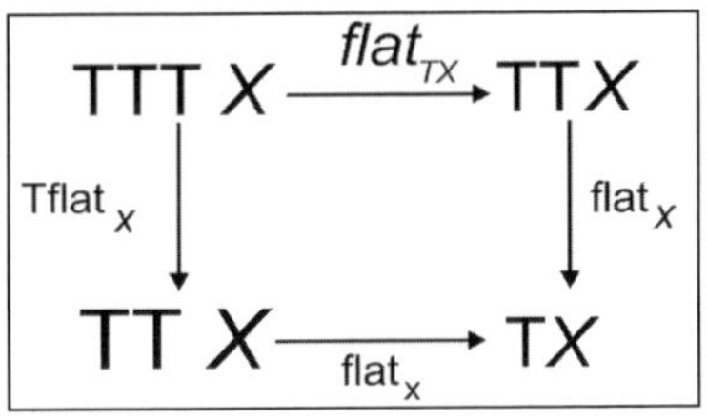

This completes the axioms of a monad, and therefore also the definition of a monad.

Eilenberg-Moore Algebras

The point of monads is that each monad comes with a natural notion of algebra, called its Eilenberg-Moore algebra. For example, in the monad of nonempty finite words, Eilenberg-Moore algebras will be semigroups, in the monad of possibly empty finite words (with the obvious definition), the Eilenberg-Moore algebras will be monoids.

Eilenberg-Moore algebra for a monad T is defined to be a pair of:

- A set A, called the universe of the algebra, and
- A multiplication operation mult: TA → A

such that multn o unit $_A$ is the identity on A, and which is associative in the sense that the following diagram commutes.

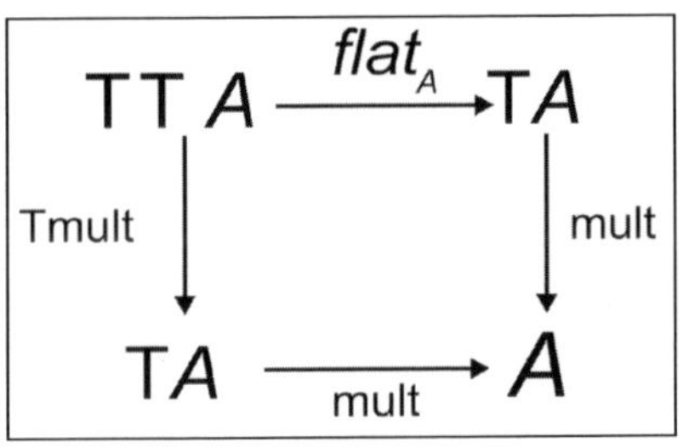

Instead of Eilenberg-Moore algebra for a monad T we will write T-algebra. We denote T-algebras by blackboard letters $\mathbb{A}$, and use the convention that the universe of $\mathbb{A}$ (respectively $\mathbb{B}$) is written by a non-boldface letter A (respectively B), while its multiplication operation is denoted by $mult_{\mathbb{A}}$ (respectively $mult_{\mathbb{B}}$). In the special case of the monad of possibly finite words, Eilenberg-Moore algebra is the same thing as a monoid; Likewise, semigroups are Eilenberg-Moore algebras for the monad of nonempty finite words.

Homomorphisms

A homomorphism of T-algebras $\mathbb{A}$ and $\mathbb{B}$ is a function $h : A \to B$ between their universes which is consistent with the multiplication operation in the sense that the following diagram commutes.

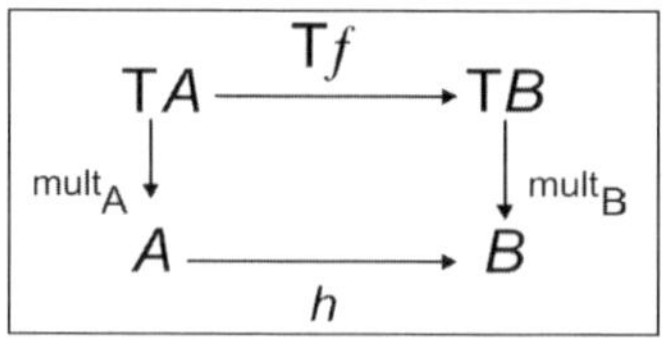

Free Algebra

For a set X, define its free T-algebra as follows: the universe is TX, while the multiplication operation is flattening. The two monoid axioms of a monad imply that this is indeed a T-algebra. By abuse of notation, we denote the free \monad-algebra by T X. For example, in the special case of the monad of non-empty finite words, the free algebra is X^+. The reason for the name free is given in the following lemma.

Free Algebra Lemma

If $\mathbb{A}$ is a T-algebra and Σ is a set, then for every function f from Σ to the universe of $\mathbb{A}$ there is a unique homomorphism of \monad-algebras h: $T\Sigma \to \mathbb{A}$ which extends it in the sense that the following diagram commutes.

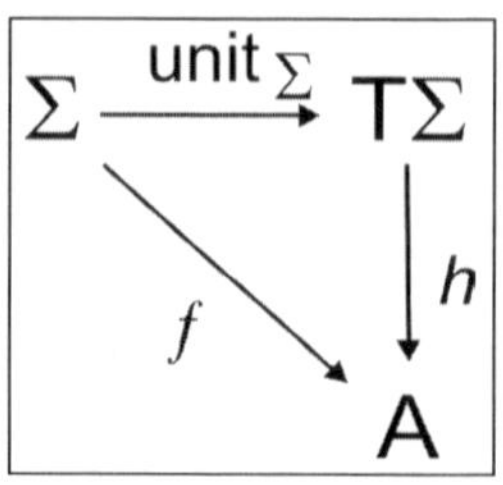

Proof: Let us first show that if the homomorphism h exists, then it is unique. Suppose that h is a homomorphism which extends f in the sense of the diagram above. Consider the following diagram.

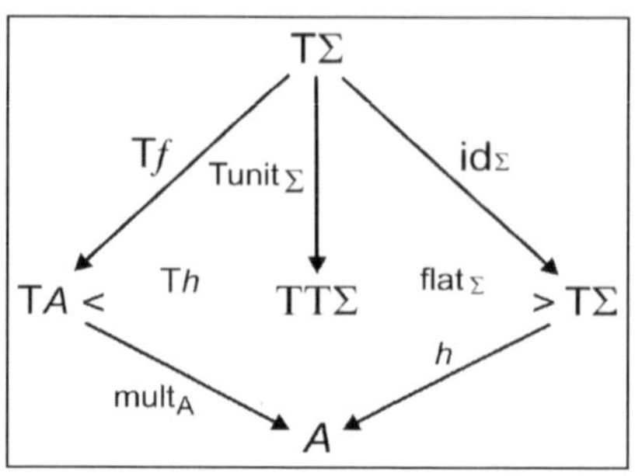

The left triangular face commutes by applying the onad, consists of the following ingreT to the diagram in the statement of the lemma which says that h extends f. (Applying a functor preserves commuting diagrams.) The right triangular face commutes by the first monoid axiom of a monad. The lower face commutes by the assumption that h is a homomorphism. Therefore, the perimeter of the diagram says that h must be equal to T f followed by multiplication in $\mathbb{A}$. We now show that h defined this way really is a homomorphism.

Being a homomorphism is that the perimeter of the following diagram commutes.

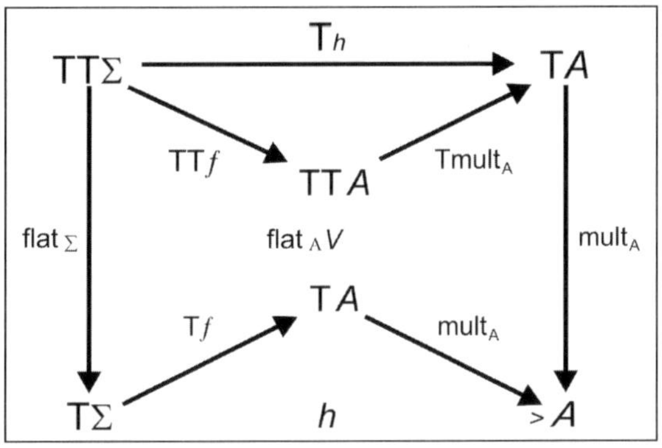

The lower triangular face is simply the definition of h, while the upper triangular face is the definition of h with the functor T applied to it. The left trapezoid face is the second of the "natural axioms", while the right trapezoid face is the associativity of multiplication in $\mathbb{A}$.

Groups

A group G is a finite or infinite set of elements together with a binary operation (called the group operation) that together satisfy the four fundamental properties of closure, associativity, the identity property, and the inverse property. The operation with respect to which a group is defined is often called the "group operation," and a set is said to be a group "under" this operation. Elements A, B, C, ... with binary operation between A and B denoted AB form a group if;

1. Closure: If A and B are two elements in G, then the product AB is also in G.
2. Associativity: The defined multiplication is associative, i.e., for all A, B, C $\in$ G, (AB) C = A(BC).
3. Identity: There is an identity element I (a.k.a. 1, E, or e) such that IA = AI = A for every element A $\in$ G.
4. Inverse: There must be an inverse (a.k.a. reciprocal) of each element. Therefore, for each element A of G, the set contains an element $B=A^{-1}$ such that $AA^{-1} = A^{-1} A = I$.

A group is a monoid each of whose elements is invertible.

A group must contain at least one element, with the unique (up to isomorphism) single-element group known as the trivial group.

The study of groups is known as group theory. If there are a finite number of elements, the group is called a finite group and the number of elements is called the group order of the group. A subset of a group that is closed under the group operation and the inverse operation is called a subgroup. Subgroups are also groups, and many commonly encountered groups are in fact special subgroups of some more general larger group.

A basic example of a finite group is the symmetric group S_n, which is the group of permutations (or "under permutation") of n objects. The simplest infinite group is the set of integers under usual addition. For continuous groups, one can consider the real numbers or the set of n × n invertible matrices. These last two are examples of Lie groups.

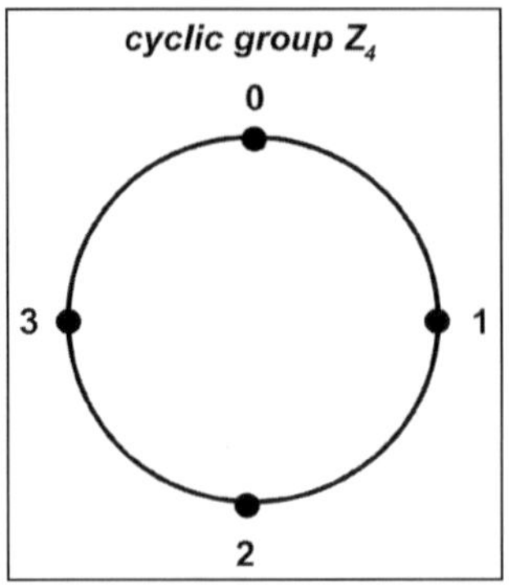

One very common type of group is the cyclic groups. This group is isomorphic to the group of integers (modulo n), is denoted Z_n, $\mathbb{Z}_n$, or $\mathbb{Z}/n\mathbb{Z}$, and is defined for every integer n > 1. It is closed under addition, associative, and has unique inverses. The numbers from 0 to n – 1 represent its elements, with the identity element represented by 0, and the inverse of i is represented by n – i.

A map between two groups which preserves the identity and the group operation is called a homomorphism. If a homomorphism has an inverse which is also a homomorphism, then it is called an isomorphism and the two groups are called isomorphic. Two groups which are isomorphic to each other are considered to be "the same" when viewed as abstract groups. For example, the group of rotations of a square, illustrated below, is the cyclic group $\mathbb{Z}_4$.

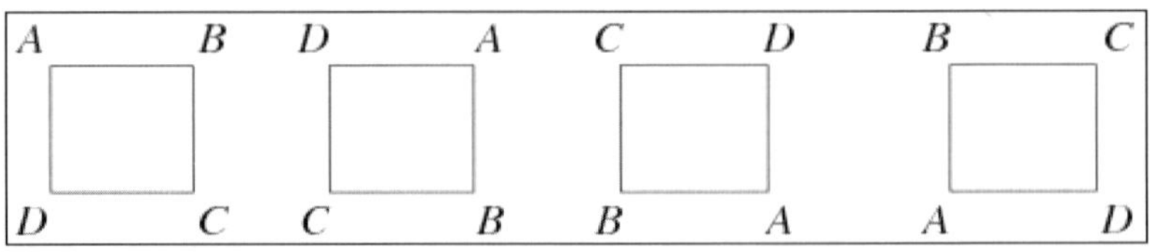

In general, a group action is when a group acts on a set, permuting its elements, so that the map from the group to the permutation group of the set is a homomorphism. For example, the rotations of a square are a subgroup of the permutations of its corners. One important group action for any group G is its action on itself by conjugation. These are just some of the possible group automorphisms. Another important kind of group action is a group representation, where the group acts on a vector space by invertible linear maps. When the field of the vector space is the complex numbers, sometimes a representation is called a CG module.

Group actions, and in particular representations, are very important in applications, not only to group theory, but also to physics and chemistry. Since a group can be thought of as an abstract mathematical object, the same group may arise in different contexts. It is therefore useful to think of a representation of the group as one particular incarnation of the group, which may also have other representations. An irreducible representation of a group is a representation for which there exists no unitary transformation which will transform the representation matrix into block diagonal form. The irreducible representations have a number of remarkable properties, as formalized in the group orthogonality theorem.

Quasi-group

A set with a binary operation (usually called multiplication) in which each of the equations ax = b and ya = b has a unique solution, for any elements a, b of the set. A quasi-group with a unit is called a loop.

A quasi-group is a natural generalization of the concept of a group. Quasi-groups arise in various areas of mathematics, for example in the theory of projective planes, non-associative division rings, in a number of questions in combinatorial analysis, etc. The term "quasi-group" was introduced by R. Moufang; it was after her work on non-Desarguesian planes, in which she elucidated the connection of such planes with quasi-groups that the development of the theory of quasi-groups properly began.

The mappings $R_\alpha : x \to xa$ and $L_\alpha : x \to ax$ are called right and left translations (or displacements) by the element a. In a quasi-group, translations are permutations of the underlying set. The subgroup G of the group of permutations of the set Q generated by all translations of the quasi-group Q(.) is called the group associated with the quasi-group Q(.). There is a close relation between the structure of G and that of Q(.).

A homomorphic image of a quasi-group need not, in general, be a quasi-group, but it is a groupoid with division. Corresponding to homomorphisms of a quasi-group onto a quasi-group are the so-called normal congruences. (A congruence) θ on Q(.) is normal if each of the relations $ac\ \theta\ bc$ and $ca\ \theta\ cb$ implies $a\ \theta\ b$.) In groups all congruences are normal. A sub-quasi-group H is called normal if there exists a normal congruence θ such that H coincides with one of the congruence classes. There exist congruences on quasi-groups in which two or even all congruence classes with respect to θ are sub-quasi-groups.

Associated with every quasi-group operation on a set are two further operations, called the left and right division operations, and denoted respectively by / and \. They are defined as follows: z/y = z and z\ x = y if x. y = z. By considering all possible permutations of x, y, z, one can obtain five further quasi-group operations apart from the initial one. Transition from the basic operation to one of the new ones is called parastrophy. Quasi-groups in which all these operations are the same as the basic one are called totally-symmetric or TS-quasi-groups. TS-quasi-groups can also be defined as quasi-groups satisfying the identities xy = yx and x (xy) = y. Idempotent TS-quasi-groups (that is, those satisfying the additional relation) $x^2 = x$ are called Steiner quasi-groups. They are closely related to Steiner triple systems.

One of the most important concepts in the theory of quasi-groups is that of an isotopy. Two quasi-groups Q and Q1 are isotopic if there exist three bijections $\rho, \sigma, \tau : Q \to Q_1$ such that $(xy)^\rho = x^\sigma y^T$

for all $x, y \in Q$. The number of non-isotopic quasi-groups that can be defined on a finite set of cardinality n is known only for $n \leq 8$.

Basic Classes of Quasi-groups

Quasi-groups are related to generalizations of groups in which the requirement of associativity is replaced by weaker conditions, now called the "A" and "B" Sushkevich postulates. A quasi-group satisfies the "A" Sushkevich postulate if the solution x of the equation $(ab)c = a(bx)$ depends only on b and c, and the "B" postulate if the solution depends only on c. It has been proved that the quasi-groups of these classes are isotopic to groups. In case the solution of such an equation depends on a and c, the quasi-group is called a left F-quasi-group. A right F-quasi-group is similarly defined via the equation $(ab)c = x(bc)$. A quasi-group that is both a left and a right F-quasi-group is called an F-quasi-group. An idempotent F-quasi-group is called a distributive quasi-group and can be defined by the identities:

$$(yz)x = (yz)(zx), \quad x(yz) = (xy)(xz),$$

called the distributive identities. It has been proved that distributive quasi-groups are isotopic to Moufang loops. Not all quasi-groups are isotopic to groups. A quasi-group Q(.) is medial if the following identity holds:

$$(xy)(uv) = (xu)(yv).$$

Every medial quasi-group is isotopic to an Abelian group Q(.) and the isotopy has the form

$$x.y = \phi x + \psi y + c,$$

where ϕ, ψ are commuting automorphisms of the group and c is some element of Q (Toyoda's theorem).

Systems of Quasi-groups and Functional Equations

Suppose that a system of quasi-groups is defined on a set Q. In this case the operations are more conveniently denoted by letters: e.g., instead of ab = c one writes $A(a,b) = c$. The quasi-group operations on Q are assumed to be related in some way, most often by identities, called in this case "functional equations". One can usually solve the problem of finding a system of quasi-groups on Q satisfying given functional equations. For example, the equation of general associativity;

$$A_1[A_2(x,y),z] = A_3[x, A_4(y,z)]$$

has been solved; namely, it has proved that if four quasi-groups satisfy (above equation), then they are isotopic to one and the same group Q(.), and the general solution is given by the equalities:

$$A_1(x,y) = \alpha x.\beta y, \quad A_2(x,y) = \alpha^{-1}(\phi x.\psi y),$$
$$A_3(x,y) = \phi x.\theta y, \quad A_4(x,y) = \theta^{-1}(\psi x.\beta y),$$

where $\alpha, \beta, \phi, \psi, \theta$ are arbitrary permutations of Q. The equation of general mediality is similarly solved:

$$A_1\left[A_2(x,y), A_3(u,v)\right] = A_4\left[A_5(x,u), A_6(y,v)\right].$$

Here all six quasi-groups turn out to be isotopic to one and the same Abelian group.

n-quasi-groups

A set with an n-ary operation is called an n-quasi-group if each of the equations:

$$\alpha_1 .....a_{i-1} x a_{i+1} .....a_n = b$$

(where $b, a_1, ..., a_n \in Q, i = 1, ..., n,$) has a unique solution. The basic concepts (isotopy, parastrophy, etc.) of the theory of quasi-groups carry over to n-quasi-groups. Each n-quasi-group is isotopic to a certain n-loop.

Certain classes of ordinary binary quasi-groups (such as the classes of medial and TS-quasi-groups, etc.) have an analogue in the n-ary case. An n-ary operation A is reducible if there exist two operations B, C, of arity at least 2, such that:

$$A(x_1, ...x_n) = B\left(x_1, ..., x_{i-1}, C\left(x_i, ..., x_j\right), x_{j+1}, ..., x_n\right)$$

(written $A = B \overset{i}{+} C$ for short). Otherwise A is said to be irreducible. An analogue of the theorem on the canonical factorization of positive integers into prime numbers holds for n-quasi-groups.

Rings

An algebraic structure (R,+,×) where R is a non-empty set and + and × are defined operations in R is called a ring if for all a, b, c in R, the following axioms are satisfied:

R_1: (R, +) is an abelian group.

1. a + b ∈ R [Closure Law for Addition]
2. (a + b) + c = a + (b +c) [Associative Law of Addition]
3. a + 0 = a = 0 + a ∀ a ∈ R [Existence of Additive Identity]
4. a + (−a) = − a + a = 0 ∀ a ∈ R [Existence of Additive Inverse]
5. a + b = b + a ∀ a ∈ R [Commutative Law of Addition]

R_2: (R, ×) is a semi group.

1. (i) a · b ∈ R [Closure Law for Multiplication]
2. (ii) (a · b) · c = a·(b · c) [Associative Law of Multiplication]

R_3: Multiplication is left as well as right distributive over addition, i.e.

$$a \cdot (b + c) = a \cdot b + a \cdot c \text{ and } (b + c) \cdot a = b \cdot a + c \cdot a$$

Elementary Properties of Rings

Some basic elementary properties of a ring can be illustrated with the help of the following theorem, and these properties are used to further develop and build concepts on rings.

Theorem: If R is a ring, then for all a, b are in R.

a. $a \cdot o = o \cdot a = a$

b. $a(-b) = (-a)b = -(ab)$

c. $(-a)(-b) = ab$

Proof:

a. We know that:

$ao = a(o + o) = ao + ao \; \forall \; a \in R$ [using distributive law].

Since R is a group under addition, applying the right cancellation law,

$$ao = ao + ao \Rightarrow a + ao = ao + ao \Rightarrow ao = o$$

Similarly, $oa = (o + o)a = oa + oa \; \forall \; a \in R$ [using distributive law].

$\therefore o + oa = oa + oa$ [becauseo $= oa + oa$]

Applying right cancellation law for addition, we get $o = oa$ i.e. $oa = o$

Thus,

$$ao = oa = o$$

b. To prove that $a(-b) = -ab$ we should show that $ab = a(-b) = o$.

We know that $a[b + (b)] = ao = o$ because $b + (-b) = o$ with the above result (a)

$ab + a(-b) = o$ [by distributive law]

$\therefore a(-b) = -(ab)$

Similarly, to show $(-a)b = -ab$, we must show that $ab + (-a)b = o$.

But, $ab + (-a)b = [a + (-a)]b = ob = o$.

$\therefore -(a)b = -(ab)$ hence the result.

c. Proving $(-a)(-b) = ab$ is a special case. However its proof is given as:

$(-a)(-b) = -[a(-b)] = [-(ab)] = ab$

This is because $-(-x) = x$ is a consequence of the fact that in a group, the inverse of the inverse of an element is the element itself.

Special Types of Rings

Commutative Rings

A ring R is said to be a commutative if the multiplication composition in R is commutative, i.e.

$ab = ba \ \forall\ a, b \in R$

Rings with Unit Element

A ring R is said to be a ring with unit element if R has a multiplicative identity, i.e. if there exists an element R denoted by 1, such that:

$$1 \cdot a = a \cdot 1 \ \forall\ a \in R$$

The ring of all $n \times n$ matrices with element as integers (rational, real or complex numbers) is a ring with unity. The unity matrix:

$$In = \begin{bmatrix} 1 & 0 & 0 & \cdots & 0 \\ 0 & 1 & 0 & \cdots & 0 \\ 0 & 0 & 1 & \cdots & 0 \\ \vdots & \vdots & \vdots & \ddots & \vdots \\ 0 & 0 & 0 & \cdots & 1 \end{bmatrix}$$

is the unity element of the ring.

Rings with or without Zero Divisors

While dealing with an arbitrary ring R, we may find elements a and b in R, where neither of which is zero and their product may be zero. We call such elements divisors of zero or zero divisors.

A ring element $a(\neq 0)$ is called a divisor of zero if there exists an element $b(\neq 0)$ in the ring such that either,

$$ab = 0 \text{ or } ba = 0$$

We also say that a ring R is without zero divisors if the product of no two non-zero elements of the same is zero, i.e. if,

$$ab = 0 \Rightarrow \text{either } a = 0 \text{ or } b = 0 \text{ or both } a = 0 \text{ and } b = 0$$

Cancellation Laws in a Ring

We say that cancellation laws hold in a ring R if,

$$ab = bc \ (a \neq 0) \Rightarrow b = c$$

$$ba = ca \ (a \neq 0) \Rightarrow b = c$$

where a, b, c are in R.

Thus in a ring with zero divisors, it is impossible to define a cancellation law.

Theorem: A ring has no divisor of zero if and only if the cancellation law holds in R.

Proof: Suppose that R has no zero divisors. Let a, b, c be any three elements of R such that $a \neq 0$, $ab = ac$.

Now,

$$ab = ac \Rightarrow ab - ac = 0$$

$$\Rightarrow a(b - c) = 0$$

$$\Rightarrow b - c = 0$$

[Because R is without zero divisors and $a \neq 0$]

$$\Rightarrow b = c$$

Thus the left cancellation law holds in R. Similarly, it can be shown that the right cancellation law also holds in R.

Conversely, suppose that the cancellation law holds in R. Let $a, b \in R$ and if possible let $ab = 0$ with $a \neq 0$, $b \neq 0$ then $ab = a \cdot 0$ (because $a \cdot 0 = 0$).

Since $a \neq 0$, $ab = a \cdot 0 \Rightarrow b = 0$.

Hence we get a contradiction to our assumption that $b \neq 0$ and therefore the theorem is established.

Division Ring

A ring is called a division ring if its non-zero elements form a group under the operation of multiplication.

Pseudo Ring

A non-empty set R with binary operations + and × satisfying all the postulates of a ring except right and left distribution laws is called pseudo ring if,

$$(a + b) \cdot (c + d) = a \cdot c + a \cdot d + b \cdot c + b \cdot d$$

for all $a, b, c, d \in R$.

Sub-rings

Let R be a ring. A non–empty subset S of the set R is said to be a subring of R if S is closed under addition and multiplication in R and S itself is a ring for those operations.

If R is any ring, then {0} and R are always subrings of R. These are said to be improper subrings. The subrings of R other than these two, if any, are said to be proper subrings of R.

Theorem: The necessary and sufficient conditions for a non-empty subset S of a ring R to be a subring of R are:

1. $a, b \in S \Rightarrow a - b \in S$
2. $a, b \in S \Rightarrow a\,b \in S$

Proof: To prove that the conditions are necessary let us suppose that S is a subring of R. Obviously S is a group with respect to addition, therefore $b \in S \Rightarrow -b \in S$.

Since S is closed under addition,

$$a \in S, b \in S \Rightarrow a \in S, -b \in S \Rightarrow a + (-b) \in S \Rightarrow a - b \in S$$

Also S is closed with respect to multiplication,

$$a \in S, b \in S \Rightarrow a\,b \in S$$

Now to prove that the conditions are sufficient, suppose S is a non-empty subset of R for which the conditions (i) and (ii) are satisfied.

From condition (i):

$$a \in S \Rightarrow a - a \in S \Rightarrow 0 \in S$$

Hence additive identity is in S. Now,

$$0 \in S, a \in S \Rightarrow 0 - a \in S \Rightarrow -a \in S$$

i.e. each element of S possesses additive inverse.

Let $a, b \in S$ then $-b \in S$ and then from condition (i):

$$0 \in S, -b \in S \Rightarrow a - (-b) \in S \Rightarrow (a + b) \in S$$

Thus S is closed under addition, and S being a subset of R, associative and commutative laws of multiplication over addition holds in S. Thus S is a subring of R.

Intersection of Subrings

Theorem: The intersection of two subrings is a subring.

Proof: Let S_1 and S_2 be two subrings of ring R.

Since, $0 \in S_1$ and $0 \in S_2$ at least $0 \in S_1 \cap S_2$. Therefore $S_1 \cap S_2$ is non-empty.

Let $a, b \in S_1 \cap S_2$, then:

$$a \in S_1 \cap S_2 \Rightarrow a \in S_1 \text{ and } a \in S_2$$

and

$$b \in S_1 \cap S_2 \Rightarrow b \in S_1 \text{ and } b \in S_2$$

But, S_1 and S_2 are subrings of R, therefore:

$$a, b \in S_1 \Rightarrow a - b \in S_1 \text{ and } ab \in S_1$$

and

$$a, b \in S_2 \Rightarrow a - b \in S_2 \text{ and } a\,b \in S_2$$

Consequently,

$$a, b \in S_1 \cap S_2 \Rightarrow a - b \in S_1 \cap S_2 \text{ and } ab \in S_1 \cap S_2.$$

Hence, $S_1 \cap S_2$ is a subring of R.

Ideals in Rings

Ideals: Let (R, +, ×) be any ring and S be a subring of R, then S is said to be a right ideal of R if $a \in S, b \in S \Rightarrow a\,b \in S$ and a left ideal of R if $a \in S, b \in S \Rightarrow ba \in S$.

Thus a non-empty subset S of R is said to be an ideal of R if:

1. S is a subgroup of R under addition.
2. For all $a \in S$ and $b \in R$, both ab and ba $\in S$.

1. Proper Ideals: If R is a commutative ring with unity and $a \in R$, the ideal $\{ax : x \in R\}$ is called the principal ideal generated by a and is denoted by (a). Thus (a) stands for the ideal generated by a.

2. Principal Ideal Ring: A commutative ring with unity for which every ideal is a principal ideal is said to be a principal ideal ring.

3. Prime Ideal: Let R be a commutative ring. An ideal P of ring R is said to be a prime ideal of R if:

$$ab \in P, a, b \in R \Rightarrow a \in P \text{ or } b \in P$$

Example: In the commutative ring of integers I, the ideal $P = \{5r : r \in I\}$ is a prime ideal since if $ab \in P$, then $5|ab$ and consequently $5|a$ or $5|b$ as 5 is prime.

Integral Domain in Rings

Integral Domain: A commutative ring with unity is said to be an integral domain if it has no zero-divisors. Alternatively a commutative ring R with unity is called an integral domain if for all $a, b \in R$, $ab = 0 \Rightarrow a = 0$ or $b = 0$.

Examples:

1. The set I of integers under usual addition and multiplication is an integral domain for any two integers a, b, $ab = 0 \Rightarrow a = 0$ or $b = 0$.

2. Consider a ring R = {0, 1, 2, 3, 4, 5, 6, 7} under the addition and multiplication modulo 8. This ring is commutative but it is not an integral domain because $2 \in R$, $4 \in R$ are two non-zero elements such that $2 \cdot 4 \equiv 0 (\text{mod} 8)$.

3. The ring of complex numbers $\mathbb{C}$ is an integral domain.

Let $J(i) = \{a + ib : a, b \in I\}$. It is easy to prove that J(i) is a commutative ring with unity. The zero element is 0+0i and unit element is 1 + 0i. Also this ring is free from zero-divisor because the product of two non-zero complex numbers cannot be zero. Hence J(i) is an integral domain.

Euclidean Ring

An integral domain R is said to be a Euclidean ring if for every $a \neq 0$ in R there is defined a non-negative integer, to be denoted by d(a), such that:

1. For all $a, b \in R$, both non-zero, $d(a) \leqslant d(ab)$,

2. For any $a, b \in R$, both non-zero, there exist $q, r \in R$ such that $a = qb + r$ when either $r = 0$ or $d(r) < d(b)$.

The set of integers $\mathbb{Z}$ depends on the property of division algorithm. This property is also known as the Euclidean algorithm, which is used to find the greatest common divisors. This property is mostly satisfied for rings, and as such we can say that such types of rings are called Euclidean rings.

Field

A field is any set of elements that satisfies the field axioms for both addition and multiplication and is commutative division algebra. An archaic name for a field is rational domain. The French term for a field is corps and the German word is Körper, both meaning "body." A field with a finite number of members is known as a finite field or Galois field.

Because the identity condition is generally required to be different for addition and multiplication, every field must have at least two elements. Examples include the complex numbers ($\mathbb{C}$), rational numbers ($\mathbb{Q}$), and real numbers ($\mathbb{R}$), but not the integers ($\mathbb{Z}$), which form only a ring.

It has been proven by Hilbert and Weierstrass that all generalizations of the field concept to triplets of elements are equivalent to the field of complex numbers.

Field Axioms

The field axioms are generally written in additive and multiplicative pairs.

Name	Addition	Multiplication
Associativity	$(a + b) + c = a (b + c)$	$(a b) c = a (b c)$
Commutativity	$a + b = b + a$	$a b = b a$
Distributivity	$a (b + c) = a b + a c$	$(a + b) c = a c + b c$
Identity	$a + 0 = a = 0 + a$	$a \cdot 1 = a = 1 \cdot a$
Inverses	$a + (-a) = 0 = (-a) + a$	$a a^{-1} = 1 = a^{-1} a$ if $a \neq 0$

References

- Abstract-algebra, wiki: brilliant.org, Retrieved 26 July, 2019
- Algebraic-structure-and-properties-of-structure, basics: includehelp.com, Retrieved 21 May, 2019
- Semi-group: encyclopediaofmath.org, Retrieved 8 January, 2019
- 1-introduction-to-monoids, alg, ~bojan: mimuw.edu.pl, Retrieved 13 May, 2019
- Introduction-to-rings-in-algebra, group-theory, tutorials: emathzone.com, Retrieved 16 January, 2019
- Special-types-of-rings, group-theory: emathzone.com, Retrieved 25 February, 2019
- Field: mathworld.wolfram.com, Retrieved 29 March, 2019

3

Linear Algebra

The domain of algebra which deals with the study of linear equations and linear functions along with their representations through vector spaces and matrices is termed linear algebra. This chapter closely examines the key concepts of linear algebra such as determinant of a matrix, transpose of a matrix and inverse of a matrix to provide an extensive understanding of the subject.

Linear algebra is the study of linear sets of equations and their transformation properties. Linear algebra allows the analysis of rotations in space, least squares fitting, solution of coupled differential equations, determination of a circle passing through three given points, as well as many other problems in mathematics, physics, and engineering. Confusingly, linear algebra is not actually algebra in the technical sense of the word "algebra" (i.e., a vector space V over a field F, and so on).

The matrix and determinant are extremely useful tools of linear algebra. One central problem of linear algebra is the solution of the matrix equation:

$$Ax = b$$

for x. While this can, in theory, be solved using a matrix inverse;

$$x = A^{-1} b,$$

other techniques such as Gaussian elimination are numerically more robust.

In addition to being used to describe the study of linear sets of equations, the term "linear algebra" is also used to describe a particular type of algebra. In particular, a linear algebra L over a field F has the structure of a ring with all the usual axioms for an inner addition and an inner multiplication together with distributive laws, therefore giving it more structure than a ring. A linear algebra also admits an outer operation of multiplication by scalars (that are elements of the underlying field F). For example, the set of all linear transformations from a vector space V to itself over a field F forms a linear algebra over F. Another example of a linear algebra is the set of all real square matrices over the field $\mathbb{R}$ of the real numbers.

Set

An algebraic set is the locus of zeros of a collection of polynomials. For example, the circle is the set of zeros of $x^2 + y^2 - 1$ and the point at (0, 0) is the set of zeros of x and y. The algebraic set $\{(x, 0)\} \cup \{(0, y)\}$ is the set of solutions to $xy = 0$. It decomposes into two irreducible algebraic sets, called algebraic varieties. In general, an algebraic set can be written uniquely as the finite union of algebraic varieties.

The intersection of two algebraic sets is an algebraic set corresponding to the union of the polynomials. For example, x=0 and y=0 intersect at (0, 0), i.e., where x = 0 and y = 0. In fact, the intersection of an arbitrary number of algebraic sets is itself an algebraic set. However, only a finite union of algebraic sets is algebraic. If X is the set of solutions to $f_i = 0$ and Y is the set of solutions to $g_j = 0$, then $X \cup Y$ is the set of solutions to $f_i g_j = 0$. Consequently, the algebraic sets are the closed sets in a topology, called the Zariski topology.

The set of polynomials vanishing on an algebraic set X is an ideal in the polynomial ring. Conversely, any ideal defines an algebraic set since it is a collection of polynomials. Hilbert's Nullstellensatz describes the precise relationship between ideals and algebraic sets.

Operations on Sets

In set theory, the operations of the sets are carried when two or more sets combined to form a single set under some of the given conditions. The basic operations on sets are:

- Union of sets.
- Intersection of sets.
- A complement of a set.
- Cartesian product of sets.
- Set difference.

Sets Formulas

Some of the most important set formulas are:

For any three sets A, B and C:

$$n(A \cup B) = n(A) + n(B) - n(A \cap B)$$

If,

$$A \cap B = \emptyset, \text{ then } n(A \cup B) - n(A) + n(B)$$

$$n(A - B) + n(A \cap B) - n(A)$$

$$n(B - A) + n(A \cap B) - n(A)$$

$$n(A - B) + n(A \cap B) + n(B - A) = n(A \cup B)$$

$$n(A \cup B \cup C) = n(A) + n(B) + n(C) - n(A \cap B) - n(B \cap C) - n(C \cap A) + n(A \cap B \cap C)$$

Representation of Sets

The sets are represented in curly braces, {}. For example, {2, 3, 4} or {a, b, c} or {Bat, Ball, Wickets}. The elements in the sets are depicted in either the Statement form, Roster Form or Set Builder Form.

Statement Form

In statement form, the well-defined descriptions of a member of a set are written and enclosed in the curly brackets.

For example, the set of even numbers less than 15.

In statement form, it can be written as {even numbers less than 15}.

Roster Form

In Roster form, all the elements of a set are listed.

For example, the set of natural numbers less than 5.

Natural Number = 1, 2, 3, 4, 5, 6, 7, 8,..........

Natural Number less than 5 = 1, 2, 3, 4

Therefore, the set is N = { 1, 2, 3, 4 }.

Set Builder Form

The general form is, A = { x : property }

For example: Write the following sets in set builder form: A={2, 4, 6, 8}

Solution:

$2 = 2 \times 1$

$4 = 2 \times 2$

$6 = 2 \times 3$

$8 = 2 \times 4$

So, the set builder form is $A = \{x : x = 2n, n \in N \text{ and } 1 \leq n \leq 4\}$.

Also, Venn Diagrams are the simple and best way for visualized representation of sets.

Types of Sets

- Empty Set: A set which does not contain any element is called an empty set or void set or null set. It is denoted by { } or Ø.
- Singleton Set: A set which contains a single element is called singleton set.
- Finite set: A set which consists of a definite number of elements is called finite set.
- Infinite set: A set which is not finite is called infinite set.
- Equivalent set: If the cardinal numbers of the two finite sets are equal, then it is called an equivalent set. I.e, n(A) = n(B).

- Equal sets: The two sets A and B are said to be equal if they have exactly the same elements.
- Subsets: A set 'A' is said to be a subset of B if every element of A is also an element of B. Intervals are subsets of R.
- Disjoint Sets: The two sets A and B are said to be disjoint if the set does not contain any common element.
- Proper set: If $A \subseteq B$ and $A \neq B$, then A is called the proper set of B and it can be written as $A \subset B$.

Elements of a Set

Let us take an example: $A = \{1, 2, 3, 4, 5\}$

Since a set is usually represented by the capital letter. Here A is the set and 1, 2, 3, 4, 5 are the elements of the set or a member of a set. The elements that are written in the set are in any order and it cannot be repeated. All the set elements are represented in small letter in case of alphabets. Also, we can write it as $1 \in A$, $2 \in A$ etc. The cardinal number of the set is 5. Some commonly used sets are as follows:

- N: Set of all natural numbers
- Z: Set of all integers
- Q: Set of all rational numbers
- R: Set of all real numbers
- Z+: Set of all positive integers

Complement of Sets

The complement of any set, say P, is the set of all elements in the universal set that are not in set P. It is denoted by P'.

Properties of Complement Sets

1. $P \cup P' = U$
2. $P \cap P' = \Phi$
3. Law of double complement : $(P')' = P$
4. Laws of empty/null set(Φ) and universal set(U), $\Phi' = U$ and $U' = \Phi$.

Properties and Laws

1. Commutative Property:
    - $A \cup B = B \cup A$
    - $A \cap B = B \cap A$

2. Associative Property:
   - $A \cup (B \cup C) = (A \cup B) \cup C$
   - $A \cap (B \cap C) = (A \cap B) \cap C$
3. Distributive Property:
   - $A \cup (B \cap C) = (A \cup B) \cap (A \cup C)$
   - $A \cap (B \cup X) = (A \cap B) \cup (A \cap X)$
4. Demorgan's Law:
   - Law of union : $(A \cup B)' = A' \cap B'$
   - Law of intersection : $(A \cap B)' = A' \cup B'$
5. Complement Law:
   - $A \cup A' = A' \cup A = U$
   - $A \cap A' = \emptyset$
6. Idempotent Law And Law of null and universal set: For any finite set A:
   - $A \cup A = A$
   - $A \cap A = A$
   - $\emptyset' = U$
   - $\emptyset = U'$

Example of Sets:

Here are few sample examples, given to represent the elements of a set.

1. Write the given statement in three methods of representation of a set:

The set of all integers that lies between -1 and 5.

Solution: The methods of representations of sets are:

- Statement Form: { I is the set of integers that lies between -1 and 5}
- Roster Form: I = { 0, 1, 2, 3, 4 }
- Set-builder Form: $I = \{ x : x \in I, -1 < x < 5 \}$

2. Find A U B and A n B and A – B. If A = {a, b, c, d} and B = {c, d}.

Solution:

- A = {a, b, c, d} and B = {c, d}

- $A \cup B = \{a, b, c, d\}$
- $A \cap B = \{c, d\}$ and
- $A - B = \{a, b\}$

Matrix

In mathematics, a matrix (plural matrices) is a rectangular array of numbers, symbols, or expressions, arranged in rows and columns. Matrices are commonly written in box brackets. The horizontal and vertical lines of entries in a matrix are called rows and columns, respectively. The size of a matrix is defined by the number of rows and columns that it contains. A matrix with m rows and n columns is called an m × n matrix or m-by-n matrix, while m and n are called its dimensions. The dimensions of the following matrix are 2 × 3 up(read "two by three"), because there are two rows and three columns.

$$A = \begin{bmatrix} 1 & 9 & -13 \\ 20 & 5 & -6 \end{bmatrix}$$

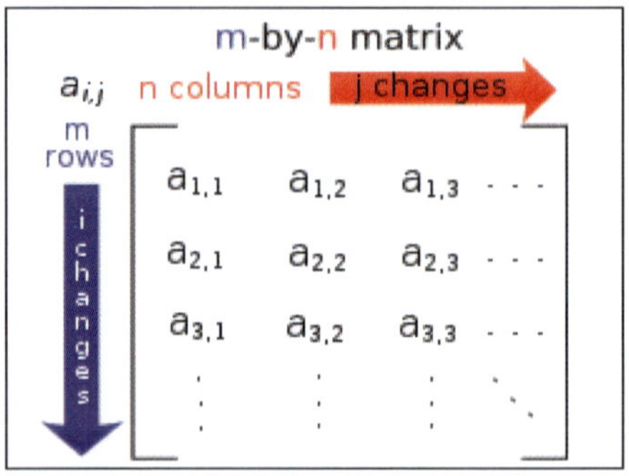

Matrix Dimensions: Each element of a matrix is often denoted by a variable with two subscripts. For instance, $a_{2,1}$ represents the element at the second row and first column of a matrix A.

The individual items (numbers, symbols or expressions) in a matrix are called its elements or entries. Provided that they are the same size (have the same number of rows and the same number of columns), two matrices can be added or subtracted element by element. The rule for matrix multiplication, however, is that two matrices can be multiplied only when the number of columns in the first equals the number of rows in the second. Any matrix can be multiplied element-wise by a scalar from its associated field.

Matrices which have a single row are called row vectors, and those which have a single column are called column vectors. A matrix which has the same number of rows and columns is called a square matrix. In some contexts, such as computer algebra programs, it is useful to consider a matrix with no rows or no columns, called an empty matrix.

Addition, Subtraction and Scalar Multiplication

Matrix addition, subtraction, and scalar multiplication are types of operations that can be applied to modify matrices.

There are a number of operations that can be applied to modify matrices, such as matrix addition, subtraction, and scalar multiplication. These form the basic techniques to work with matrices.

These techniques can be used in calculating sums, differences and products of information such as sodas that come in three different flavors: apple, orange, and strawberry and two different packaging: bottle and can. Two tables summarizing the total sales between last month and this month are written to illustrate the amounts. Matrix addition, subtraction and scalar multiplication can be used to find such things as: the sales of last month and the sales of this month, the average sales for each flavor and packaging of soda in the 2-month period.

Adding and Subtracting Matrices

We use matrices to list data or to represent systems. Because the entries are numbers, we can perform operations on matrices. We add or subtract matrices by adding or subtracting corresponding entries.

In order to do this, the entries must correspond. Therefore, addition and subtraction of matrices is only possible when the matrices have the same dimensions. Matrix addition is commutative and is also associative, so the following is true:

$$A + B = B + A$$

$$(A + B) + C = A + (B + C)$$

Adding matrices is very simple. Just add each element in the first matrix to the corresponding element in the second matrix.

$$\begin{pmatrix} 1 & 2 & 3 \\ 4 & 5 & 6 \end{pmatrix} + \begin{pmatrix} 10 & 20 & 30 \\ 40 & 50 & 60 \end{pmatrix} = \begin{pmatrix} 11 & 22 & 33 \\ 44 & 55 & 66 \end{pmatrix}$$

Note that element in the first matrix, 1, adds to element x_{11} in the second matrix, 10, to produce element x_{11} in the resultant matrix, 11. Also note that both matrices being added are 2 × 3, and the resulting matrix is also 2 × 3. You cannot add two matrices that have different dimensions.

As you might guess, subtracting works much the same way except that you subtract instead of adding.

$$\begin{pmatrix} 10 & -20 & 30 \\ 40 & 50 & 60 \end{pmatrix} - \begin{pmatrix} 1 & -2 & 3 \\ 4 & -5 & 6 \end{pmatrix} = \begin{pmatrix} 9 & -18 & 27 \\ 36 & 55 & 54 \end{pmatrix}$$

Once again, note that the resulting matrix has the same dimensions as the originals, and that you cannot subtract two matrices that have different dimensions. Be careful when subtracting with signed numbers.

Scalar Multiplication

In an intuitive geometrical context, scalar multiplication of a real Euclidean vector by a positive real number multiplies the magnitude of the vector without changing its direction. What does it mean to multiply a number by 3? It means you add the number to itself 3 times. Multiplying a matrix by 3 means the same thing; you add the matrix to itself 3 times, or simply multiply each element by that constant.

$$3.\begin{pmatrix} 1 & 2 & 3 \\ 4 & 5 & 6 \end{pmatrix} = \begin{pmatrix} 3 & 6 & 9 \\ 12 & 15 & 18 \end{pmatrix}$$

The resulting matrix has the same dimensions as the original. Scalar multiplication has the following properties:

- Left and right distributivity: (c + d) M = M (c + d) = Mc + Md
- Associativity: (cd)M = c(dM)
- Identity: 1M = M
- Null: 0M = 0
- Additive inverse: (−1)M = − M

Matrix Multiplication

When multiplying matrices, the elements of the rows in the first matrix are multiplied with corresponding columns in the second matrix.

If A is an n × m matrix and B is an m × p matrix, the result AB of their multiplication is an n × p matrix defined only if the number of columns m in A is equal to the number of rows m in B. Check to make sure that this is true before multiplying the matrices, since there is "no solution" otherwise.

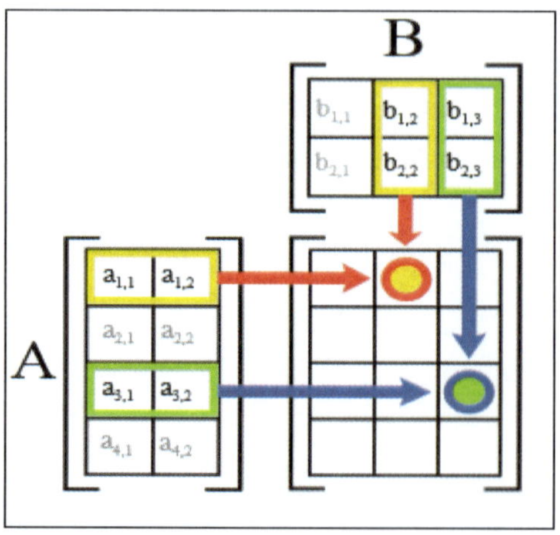

Matrix Multiplication: This figure illustrates diagrammatically the product of two matrices A and B, showing how each intersection in the product matrix corresponds to a row of A and a column of B.

Scalar multiplication is simply multiplying a value through all the elements of a matrix, whereas matrix multiplication is multiplying every element of each row of the first matrix times every element of each column in the second matrix. Scalar multiplication is much more simple than matrix multiplication; however, a pattern does exist.

When multiplying matrices, the elements of the rows in the first matrix are multiplied with corresponding columns in the second matrix. Each entry of the resultant matrix is computed one at a time.

For two matrices the final position of the product is shown below:

$$\begin{bmatrix} a_{11} & a_{12} \\ \cdot & \cdot \\ a_{31} & a_{32} \\ \cdot & \cdot \end{bmatrix} \begin{bmatrix} \cdot & b_{12} & b_{13} \\ \cdot & b_{22} & b_{23} \end{bmatrix} = \begin{bmatrix} \cdot & x_{12} & \cdot \\ \cdot & \cdot & \cdot \\ \cdot & \cdot & x_{33} \\ \cdot & \cdot & \cdot \end{bmatrix}$$

The values at the intersections marked with circles are:

$$x_{12} = (a_{11}, a_{12}) \cdot (b_{12}, b_{22}) = (a_{11}b_{12}) + (a_{12}b_{22})$$
$$x_{33} = (a_{31}, a_{32}) \cdot (b_{13}, b_{23}) = (a_{31}b_{13}) + (a_{32}b_{23})$$

Matrix Multiplication: Process

Example: Find the product AB

$$A = \begin{pmatrix} 1 & 2 \\ 3 & 4 \end{pmatrix} \quad B = \begin{pmatrix} 5 & 6 \\ 7 & 8 \end{pmatrix}$$

Do the number of columns in A equal the number of rows in B? The number of columns in A is 2, and the number of rows in B is also 2, therefore a product exists.

Start with producing the product for the first row, first column element. Take the first row of Matrix A and multiply by the first column of Matrix B: The first element of A times the first column element of B, plus the second element of A times the second column element of B.

$$AB = \begin{pmatrix} (1\cdot 5) + (2\cdot 7) & (\ ) + (\ ) \\ (\ ) + (\ ) & (\ ) + (\ ) \end{pmatrix}$$

Continue the pattern with the first row of A by the second column of B, and then repeat with the second row of A.

AB has entries defined by the equation:

$$AB = \begin{pmatrix} (1\cdot 5) + (2\cdot 7) & (1\cdot 6) + (2\cdot 8) \\ (3\cdot 5) + (4\cdot 7) & (3\cdot 6) + (4\cdot 8) \end{pmatrix}$$

$$AB = \begin{pmatrix} (5+14) & (6+16) \\ (15+28) & (18+32) \end{pmatrix}$$

$$AB = \begin{pmatrix} (19) & (22) \\ (43) & (50) \end{pmatrix}$$

Matrix Equations

Matrices can be used to compactly write and work with systems of multiple linear equations.

Matrices can be used to compactly write and work with systems of equations. matrices can be manipulated in any way that a normal equation can be. This is very helpful when we start to work with systems of equations. It is helpful to understand how to organize matrices to solve these systems.

Writing a System of Equations with Matrices

It is possible to solve this system using the elimination or substitution method, but it is also possible to do it with a matrix operation. Before we start setting up the matrices, it is important to do the following:

- Make sure that all of the equations are written in a similar manner, meaning the variables need to all be in the same order.
- Make sure that one side of the equation is only variables and their coefficients, and the other side is just constants.

Solving a system of linear equations using the inverse of a matrix requires the definition of two new matrices: X is the matrix representing the variables of the system, and B is the matrix representing the constants. Using matrix multiplication, we may define a system of equations with the same number of equations as variables as:

$$A \cdot X = B$$

To solve a system of linear equations using an inverse matrix, let A be the coefficient matrix, let

X be the variable matrix, and let B be the constant matrix.

Given the system:

$$\begin{aligned} x + 8y &= 7 \\ 2x - 8y &= -3 \end{aligned}$$

The coefficient matrix is:

$$A = \begin{bmatrix} 1 & 8 \\ 2 & -8 \end{bmatrix}$$

The variable matrix is:

$$X = \begin{bmatrix} x \\ y \end{bmatrix}$$

The constant matrix is:

$$B = \begin{bmatrix} 7 \\ -3 \end{bmatrix}$$

Thus, to solve a system AX = B, for X, multiply both sides by the inverse of A and we shall obtain the solution:

$$X = \left(A^{-1}\right)B$$

Provided the inverse (A^{-1}) exists, this formula will solve the system.

If the coefficient matrix is not invertible, the system could be inconsistent and have no solution, or be dependent and have infinitely many solutions.

Matrices and Row Operations

Two matrices are row equivalent if one can be changed to the other by a sequence of elementary row operations.

Elementary Row Operations (ERO)

In linear algebra, two matrices are row equivalent if one can be changed to the other by a sequence of elementary row operations. Alternatively, two $m \times n$ matrices are row equivalent if and only if they have the same row space. The row space of a matrix is the set of all possible linear combinations of its row vectors. If the rows of the matrix represent a system of linear equations, then the row space consists of all linear equations that can be deduced algebraically from those in the system. Two matrices of the same size are row equivalent if and only if the corresponding homogeneous systems have the same set of solutions, or equivalently the matrices have the same null space. Because elementary row operations are reversible, row equivalence is an equivalence relation. It is commonly denoted by a tilde (~).

An elementary row operation is any one of the following three moves:

1. Row switching (swap): Swap two rows of a matrix.
2. Row multiplication (scale): Multiply a row of a matrix by a nonzero constant.
3. Row addition (pivot): Add to one row of a matrix some multiple of another row.

Produce Equivalent Matrices using Elementary Row Operations

Since the matrix is essentially the coefficients and constants of a linear system, the three row operations preserve the matrix. For example, swapping two rows simply means switching their position within the matrix. Also, when solving a system of linear equations by the elimination method, row multiplication would be the same as multiplying the whole equation by a number to obtain additive inverses so that a variable cancels. Finally, row addition is also the same as the elimination method, when one chooses to add or subtract the like terms of the equations to obtain the variable. Therefore, row operations preserve the matrix and can be used as an alternative method to solve a system of equations.

Example: Show that these two matrices are row equivalent:

$$A = \begin{pmatrix} 1 & -1 & 0 \\ 2 & 1 & 1 \end{pmatrix} \quad B = \begin{pmatrix} 3 & 0 & 1 \\ 0 & 3 & 1 \end{pmatrix}$$

Start with A, add the second row to the first:

$$A = \begin{pmatrix} 3 & 0 & 1 \\ 2 & 1 & 1 \end{pmatrix}$$

Then, multiply the second row by 3 and then subtract the first row from the second:

$$A = \begin{pmatrix} 3 & 0 & 1 \\ 3 & 3 & 2 \end{pmatrix}$$

Finally, subtract the first row from the second:

$$A = \begin{pmatrix} 3 & 0 & 1 \\ 0 & 3 & 1 \end{pmatrix}$$

You can see that A = B, which we achieved through a series of elementary row operations.

Row Reduction: Solving a System of Linear Equations

In row reduction, the linear system:

$$x + 3y - 2z = 5$$

$$3x + 5y + 6z = 7$$

$$2x + 4y + 3z = 8$$

Is represented as an augmented matrix:

$$A = \begin{pmatrix} 1 & 3 & -2 & 5 \\ 3 & 5 & 6 & 7 \\ 2 & 4 & 3 & 8 \end{pmatrix}$$

This matrix is then modified using elementary row operations until it reaches reduced row echelon form.

Because these operations are reversible, the augmented matrix produced always represents a linear system that is equivalent to the original.

There are several specific algorithms to row-reduce an augmented matrix, the simplest of which are Gaussian elimination and Gauss-Jordan elimination. This computation can be done by hand (using the three types of ERO) or on the calculator under the matrix function 'rref' (reduced row echelon form).

The final matrix is in reduced row echelon form, and represents the system $x = -15, y = 8z = 2$.

$$A = \begin{pmatrix} 1 & 0 & 0 & -15 \\ 0 & 1 & 0 & 8 \\ 0 & 0 & 1 & 2 \end{pmatrix}$$

Simplifying Matrices with Row Operations

Using elementary operations, Gaussian elimination reduces matrices to row echelon form.

By means of a finite sequence of elementary row operations, called Gaussian elimination, any matrix can be transformed to a row echelon form. This transformation is necessary for solving a system of linear equations.

Before getting into more detail, there are a couple of key terms that should be mentioned:

- Augmented matrix: an augmented matrix is a matrix obtained by appending the columns of two given matrices, usually for the purpose of performing the same elementary row operations on each of the given matrices.
- Upper triangle form: A square matrix is called upper triangular if all the entries below the main diagonal are zero. A triangular matrix is one that is either lower triangular or upper triangular. A matrix that is both upper and lower triangular is a diagonal matrix.
- Elementary row operations: Swap rows, add rows or multiply rows.

Gaussian Elimination

- Write the augmented matrix for the linear equations.
- Use elementary row operations on the augmented matrix [A|b] to transform A to upper triangle form. If a zero is on the diagonal, switch the rows until a nonzero is in its place.
- Use back substitution to find the solution.

Example: Solve the system by Gaussian Elimination:

$$2x + y - z = 8$$
$$-3x - y + 2z = -11$$
$$-2x + y + 2z = -3$$

Write the augmented matrix:

$$\left[\begin{array}{ccc|c} 2 & 1 & -1 & 8 \\ -3 & -1 & 2 & -11 \\ -1 & 1 & 2 & -3 \end{array}\right]$$

Use elementary row operations to reduce the matrix to reduced row echelon form:

$$\left[\begin{array}{ccc|c} 1 & 0 & 0 & 2 \\ 0 & 1 & 0 & 3 \\ 0 & 0 & 1 & -1 \end{array}\right]$$

Using elementary row operations to obtain reduced row echelon form ('rref' in the calculator) the solution to the system is revealed in the last column: $x = 2,\ y = 3,\ z = 1$.

The Inverse of a Matrix

The inverse of matrix [A] is $[A]^{-1}$, and is defined by the property: $[A][A]^{-1} = [A]^{-1}[A] = [I]$.

Having seen that the number 1 plays a special role in multiplication, 1x = x, the inverse of a number is defined as a number that multiplies by that number to give 1. The matrix BB is the inverse of the matrix AA if when multiplied together, A·B or B·A gives the identity matrix.

The inverse of matrix [A], designated as $[A]^{-1}$, is defined by the property:

$$[A][A]^{-1} = [A]^{-1} [A] = [I]$$

Where [I] is the identity matrix.

Note that, just as in the definition of the identity matrix, this definition requires commutativity—the multiplication must work the same in either order.

Note also that only square matrices can have an inverse. The definition of an inverse matrix is based on the identity matrix [I], and it has already been established that only square matrices have an associated identity matrix.

The method for finding an inverse matrix comes directly from the definition, along with a little algebra.

Finding an Inverse Matrix

Example: Find the inverse of: $\begin{pmatrix} 3 & 4 \\ 5 & 6 \end{pmatrix}$

First, let the following be true:

$$\begin{pmatrix} 3 & 4 \\ 5 & 6 \end{pmatrix}\begin{pmatrix} a & b \\ c & d \end{pmatrix} = \begin{pmatrix} 1 & 0 \\ 0 & 1 \end{pmatrix}$$

This is the key step. It establishes $\begin{pmatrix} a & b \\ c & d \end{pmatrix}$ as the inverse that is being looked for, by asserting that it fills the definition of an inverse matrix. When multiplying this mystery matrix by our original matrix, the result is [I]. When solving for the four variables a, b, c, and d, then the inverse of the matrix will be found.

Next, do the multiplication:

$$\begin{pmatrix} 3a+4c & 3b+4d \\ 5a+6c & 5d+6d \end{pmatrix} = \begin{pmatrix} 1 & 0 \\ 0 & 1 \end{pmatrix}$$

For two matrices to be equal, every element in the left must equal its corresponding element on the right. So, for these two matrices to equal each other:

$$3a + 4c = 1$$

$$3b + 4d = 0$$

5a + 6c = 0

5b + 6d = 1

Solve the first two equations for a and c and the second two equations for b and d by using either elimination or substitution.

The results are:

$$a = -3$$

$$b = 2$$

$$c = 2\frac{1}{2}$$

$$d = -1\frac{1}{2}$$

Having solved for the four variables, the result is the inverse: $\begin{pmatrix} -3 & 2 \\ 2\frac{1}{2} & -1\frac{1}{2} \end{pmatrix}$

If an inverse has been found, then a quick check to be sure it is correct is to multiply it by the original matrix and see if the identify matrix results:

$$\begin{pmatrix} -3 & 2 \\ 2\frac{1}{2} & -1\frac{1}{2} \end{pmatrix} \begin{pmatrix} 3 & 4 \\ 5 & 6 \end{pmatrix}$$

After multiplying, the result is:

$$\begin{pmatrix} (-3)(3)+(5)(5) & (-3)(4)+(2)(6) \\ \left(2\frac{1}{2}\right)(3)+\left(-1\frac{1}{2}\right)(5) & \left(2\frac{1}{2}\right)(4)+\left(-1\frac{1}{2}\right)(6) \end{pmatrix}$$

Simplifying the problem gives:

$$\begin{pmatrix} -9+10 & -12+12 \\ 7\frac{1}{2}-7\frac{1}{2} & 10-9 \end{pmatrix} = \begin{pmatrix} 1 & 0 \\ 0 & 1 \end{pmatrix}$$

Therefore, the inverse works.

Note that, to fully test it, one has to try the multiplication in both orders, because, in general, changing the order of a matrix multiplication changes the answer. The definition of an inverse matrix specifies that it must work both ways.

$$\begin{pmatrix} 3 & 4 \\ 5 & 6 \end{pmatrix} \begin{pmatrix} -3 & 2 \\ 2\frac{1}{2} & -1\frac{1}{2} \end{pmatrix}$$

After multiplying, one gets:

$$\begin{pmatrix} (3)(-3)+(4)\left(2\frac{1}{2}\right) & (3)(2)+(4)\left(-1\frac{1}{2}\right) \\ (5)(-3)+(6)\left(2\frac{1}{2}\right) & (5)(2)+(6)\left(-1\frac{1}{2}\right) \end{pmatrix}$$

Simplifying the problem gives:

$$\begin{pmatrix} -9+10 & 6-6 \\ -15+15 & 10-9 \end{pmatrix} = \begin{pmatrix} 1 & 0 \\ 0 & 1 \end{pmatrix}$$

So the inverse matrix works both ways.

In some cases, the inverse of a square matrix does not exist. This is called a singular matrix.

Solving Systems of Equations using Matrix Inverses

A system of equations can be readily solved using the concept of the inverse matrix and matrix multiplication.

Example : Solve the following system of linear equations:

$$x+2y-z=11$$
$$2x-y+3z=7$$
$$7x-3y-2z=2$$

Step 1: Set up a matrix A, which is comprised of the coefficients of all the variables from each equation from the left hand side of the equal sign:

$$[A]=\begin{bmatrix} 1 & 2 & -1 \\ 2 & -1 & 3 \\ 7 & -3 & -2 \end{bmatrix}$$

Step 2: Step up a matrix B, which consists of the constants on the right hand side of the equal sign:

$$[B]=\begin{bmatrix} 11 \\ 7 \\ 2 \end{bmatrix}$$

Now, in order to determine the values of x, y, and z, we simply multiply the inverse of [A] times [B]. This can be done by hand, finding the inverse matrix of [A], then performing the appropriate matrix multiplication with [B].

However, if you have a graphing calculator, the situation is much easier. Using the matrix function on the calculator, first enter both matrices. Then calculate [A^{-1}][B], that is, the inverse of matrix [A], multiplied by matrix [B].

The calculator responds with a 3 × 1 matrix, which includes the solution to the system as: x = 3, y = 5, and z = 2. Solving linear equations in this way is fast and easy.

Determinants of 2-by-2 Square Matrices

The determinant of a 2 × 2 square matrix is a mathematical construct used in problem solving that is found by a special formula.

A matrix is often used to represent the coefficients in a system of linear equations, and the determinant can be used to solve those equations. The use of determinants in calculus includes the Jacobian determinant in the change of variables rule for integrals of functions of several variables. Determinants are also used to define the characteristic polynomial of a matrix, which is essential for eigenvalue problems in linear algebra. In analytical geometry, determinants express the signed n-dimensional volumes of n-dimensional parallelepipeds. Sometimes, determinants are used merely as a compact notation for expressions that would otherwise be unwieldy to write down.

It can be proven that any matrix has a unique inverse if its determinant is nonzero. Various other theorems can be proved as well, including that the determinant of a product of matrices is always equal to the product of determinants; and, the determinant of a Hermitian matrix is always real.

The determinant of a matrix [A] is denoted det (A), det A, or |A|. In the case where the matrix entries are written out in full, the determinant is denoted by surrounding the matrix entries by vertical bars instead of the brackets or parentheses of the matrix.

For instance, the determinant of the matrix $\begin{bmatrix} a & b \\ b & e \end{bmatrix}$ is written $\begin{bmatrix} a & b \\ b & e \end{bmatrix}$.

Determinant of a 2-by-2 Matrix

In linear algebra, the determinant is a value associated with a square matrix. It can be computed from the entries of the matrix by a specific arithmetic expression, shown below:

For a 2 × 2 matrix, $\begin{bmatrix} a & b \\ c & d \end{bmatrix}$, the determinant $\begin{bmatrix} a & b \\ c & d \end{bmatrix}$ is defined to be ad – bc.

Example: Find the determinant of the following matrix:

$$\begin{bmatrix} 4 & -2 \\ 7 & 5 \end{bmatrix}$$

The determinant $\begin{bmatrix} 4 & -2 \\ 7 & 5 \end{bmatrix}$ is: $(4 \cdot 5)-(-2 \cdot 7)=20-(-14)=34$.

Cofactors, Minors and Further Determinants

The cofactor of an entry (i, j) of a matrix A is the signed minor of that matrix.

Cofactor

In linear algebra, the cofactor (sometimes called adjunct) describes a particular construction that is useful for calculating both the determinant and inverse of square matrices. Specifically the cofactor of the (i, j) entry of a matrix, also known as the (i, j) cofactor of that matrix, is the signed minor of that entry.

The cofactor of aij entry of a matrix is defined as:

$$C_{ij} = (-1)^{1+j} M_{ij}$$

Minor

To know what the signed minor is, we need to know what the minor of a matrix is. In linear algebra, a minor of a matrix A is the determinant of some smaller square matrix, cut down from A by removing one or more of its rows or columns. Minors obtained by removing just one row and one column from square matrices (first minors) are required for calculating matrix cofactors.

Let A be an $m \times n$ matrix and k an integer with $0 < k \leq m$, and $k \leq n$. A $k \times k$ minor of A is the determinant of a $k \times k$ matrix obtained from A by deleting $m - k$ rows and $n - k$ columns.

Calculate the Determinant

The determinant of any matrix can be found using its signed minors. The determinant is the sum of the signed minors of any row or column of the matrix scaled by the elements in that row or column.

Calculating the Minors

The following steps are used to find the determinant of a given minor of a matrix A:

1. Choose an entry a_{ij} from the matrix.
2. Cross out the entries that lie in the corresponding row ii and column j.
3. Rewrite the matrix without the marked entries.
4. Obtain the determinant of this new matrix.

M_{ij} is termed the minor for entry a_{ij}.

If $i + j$ is an even number, the cofactor coincides with its minor: $C_{ij} = M_{ij}$. Otherwise, it is equal to the additive inverse of its minor: $C_{ij} = -M_{ij}$

Calculating the Determinant

We will find the determinant of the following matrix A by calculating the determinants of its cofactors for the third, rightmost column and then multiplying them by the elements of that column.

$$\begin{bmatrix} 1 & 4 & 7 \\ 3 & 0 & 5 \\ -1 & 9 & 11 \end{bmatrix}$$

As an example, we will calculate the determinant of the minor M_{23}, which is the determinant of the 2 × 2 matrix formed by removing the 2nd row and 3rd column. A black dot represents an element we are removing.

$$\begin{bmatrix} 1 & 4 & \bullet \\ \bullet & \bullet & \bullet \\ -1 & 9 & \bullet \end{bmatrix} = \begin{bmatrix} 1 & 4 \\ 1 & 9 \end{bmatrix}$$

$$= (9-(-4))$$

$$= 13$$

Since, I + j = 5 is an odd number, the cofactor is the additive inverse of its minor: – (13) = – 13.

We multiply this number by a_{23} = 5, giving −65.

The same process is carried out to find the determinants of C_{13} and C_{33}, which are then multiplied by a_{13} and a_{33}, respectively. The determinant is then found by summing all of these:

$$\det A = a_{13} \det C_{13} + a_{23} \det C_{23} + a_{33} \det C_{33}$$

$$= 7 \cdot 27 - 5 \cdot 13 + 11 \cdot -12$$

$$= -8$$

Cramer's Rule

Cramer's Rule uses determinants to solve for a solution to the equation A x = b, when A is a square matrix.

"Cramer's Rule" is another way to solve a system of linear equations with matrices. It uses a formula to calculate the solution to the system utilizing the definition of determinants.

Cramer's Rule is an explicit formula for the solution of a system of linear equations with as many equations as unknowns, i.e. a square matrix, valid whenever the system has a unique solution. It expresses the solution in terms of the determinants of the (square) coefficient matrix and of matrices obtained from it by replacing one column by the vector of right hand sides of the equations.

Cramer's Rule: Formula

Rules for a 2 × 2 Matrix

Consider the linear system:

$$\begin{bmatrix} a & b \\ c & d \end{bmatrix} \begin{bmatrix} x \\ y \end{bmatrix} = \begin{bmatrix} e \\ f \end{bmatrix}$$

Assume the determinant is non-zero. Then, x and y can be found by Cramer's rule:

$$x = \frac{\begin{vmatrix} e & b \\ f & d \end{vmatrix}}{\begin{vmatrix} a & b \\ c & d \end{vmatrix}} = \frac{ed - bf}{ad - bc}$$

And,

$$y = \frac{\begin{vmatrix} a & e \\ c & f \end{vmatrix}}{\begin{vmatrix} a & b \\ c & d \end{vmatrix}} = \frac{ad - ec}{ad - bc}$$

Rules for a 3 × 3 Matrix

$$\begin{bmatrix} a & b & c \\ b & e & f \\ g & h & i \end{bmatrix} \begin{bmatrix} x \\ y \\ z \end{bmatrix} = \begin{bmatrix} j \\ k \\ l \end{bmatrix}$$

Then the values of x, y and z can be found as follows:

$$x = \frac{\begin{vmatrix} j & b & c \\ k & e & f \\ l & h & i \end{vmatrix}}{\begin{vmatrix} a & b & c \\ d & e & f \\ g & h & i \end{vmatrix}} \quad y = \frac{\begin{vmatrix} a & j & c \\ d & k & f \\ g & l & i \end{vmatrix}}{\begin{vmatrix} a & b & c \\ d & e & f \\ g & h & i \end{vmatrix}} \quad z = \frac{\begin{vmatrix} a & b & j \\ d & e & k \\ g & h & l \end{vmatrix}}{\begin{vmatrix} a & b & c \\ d & e & f \\ g & h & i \end{vmatrix}}$$

Using Cramer's Rule

Example: Solve the system using Cramer's Rule:

$$\begin{cases} 3x + 2y = 10 \\ -6x + 4y = 4 \end{cases}$$

In matrix format:

$$\begin{bmatrix} 3 & 2 \\ -6 & 4 \end{bmatrix} \begin{bmatrix} x \\ y \end{bmatrix} = \begin{bmatrix} 10 \\ 4 \end{bmatrix}$$

$$x = \frac{\begin{vmatrix} e & b \\ f & d \end{vmatrix}}{\begin{vmatrix} a & b \\ c & d \end{vmatrix}}$$

$$= \frac{ed - bf}{ad - bc}$$

$$x = \frac{\begin{vmatrix} 10 & 2 \\ 4 & 4 \end{vmatrix}}{\begin{vmatrix} 3 & 2 \\ -6 & 4 \end{vmatrix}}$$

$$= \frac{10 \cdot 4 - 2 \cdot 4}{(3 \cdot 4) - [2 \cdot (-6)]}$$

$$= \frac{32}{24} = \frac{4}{3}$$

$$y = \frac{\begin{vmatrix} a & e \\ c & f \end{vmatrix}}{\begin{vmatrix} a & b \\ c & d \end{vmatrix}}$$

$$= \frac{af - ec}{ad - bc}$$

$$y = \frac{\begin{vmatrix} 3 & 10 \\ -6 & 4 \end{vmatrix}}{\begin{vmatrix} 3 & 2 \\ -6 & 4 \end{vmatrix}}$$

$$= \frac{(3 \cdot 4) - [10 \cdot (-6)]}{(3 \cdot 4) - [2 \cdot (-6)]}$$

$$= \frac{72}{24} = 3$$

The solution to the system is $\left(\frac{4}{3}, 3\right)$.

Types of Matrices

Matrices are distinguished on the basis of their order, elements and certain other conditions. There are different types of matrices but the most commonly used are discussed below. Let's find out the types of matrices in the field of mathematics.

Different types of Matrices and their forms are used for solving numerous problems. Some of them are as follows:

$$\begin{bmatrix} a_{11} & a_{12} & \cdots & a_{1n} \\ a_{21} & a_{22} & \cdots & a_{2n} \\ \vdots & \vdots & \ddots & \vdots \\ a_{m1} & a_{m2} & \cdots & a_{mn} \end{bmatrix}$$

Row Matrix

A row matrix has only one row but any number of columns. A matrix is said to be a row matrix if it has only one row. For example,

$$A = \begin{bmatrix} -1/2 & \sqrt{5} & 2 & 3 \end{bmatrix}$$

is a row matrix of order 1 × 4. In general, $A = [a_{ij}]_{1 \times n}$ is a row matrix of order 1 × n.

Column Matrix

A column matrix has only one column but any number of rows. A matrix is said to be a column matrix if it has only one column. For example,

$$A = \begin{bmatrix} 0 \\ \sqrt{3} \\ -1 \\ 1/2 \end{bmatrix}$$

is a column matrix of order 4 × 1. In general, $B = [b_{ij}]_{m \times 1}$ is a column matrix of order m × 1.

Square Matrix

A square matrix has the number of columns equal to the number of rows. A matrix in which the number of rows is equal to the number of columns is said to be a square matrix. Thus an m × n matrix is said to be a square matrix if m = n and is known as a square matrix of order 'n'. For example,

$$A = \begin{bmatrix} 3 & -1 & 0 \\ 3/2 & \sqrt{3}/2 & 1 \\ 4 & 3 & -1 \end{bmatrix}$$

is a square matrix of order 3. In general, $A = [a_{ij}]$ m × m is a square matrix of order m.

Rectangular Matrix

A matrix is said to be a rectangular matrix if the number of rows is not equal to the number of columns. For example,

$$A = \begin{bmatrix} 3 & -1 & 0 \\ 3/2 & \sqrt{3}/2 & -1 \\ 4 & 3 & -1 \\ 7/2 & 2 & -5 \end{bmatrix}$$

is a matrix of the order 4 × 3.

Diagonal Matrix

A square matrix B = [bij] m × m is said to be a diagonal matrix if all its non-diagonal elements are zero, that is a matrix $B = [b_{ij}]_{m \times m}$ is said to be a diagonal matrix if $b_{ij} = 0$, when $i \neq j$. For example,

$$A = [4] \begin{bmatrix} -1 & 0 \\ 0 & 2 \end{bmatrix} \begin{bmatrix} 3 & 0 & 0 \\ 0 & -5 & 0 \\ 0 & 0 & 2 \end{bmatrix}$$

are diagonal matrices of order 1, 2, 3, respectively.

Scalar Matrix

A diagonal matrix is said to be a scalar matrix if all the elements in its principal diagonal are equal to some non-zero constant. A diagonal matrix is said to be a scalar matrix if its diagonal elements are equal, that is, a square matrix $B = [b_{ij}]_{n \times n}$ is said to be a scalar matrix if:

- $b_{ij} = 0$, when $i \neq j$.
- $b_{ij} = k$, when $i = j$, for some constant k.

For example,

$$A = [4] \begin{bmatrix} -1 & 0 \\ 0 & -1 \end{bmatrix} \begin{bmatrix} 3 & 0 & 0 \\ 0 & 3 & 0 \\ 0 & 0 & 3 \end{bmatrix}$$

are scalar matrices of order 1, 2 and 3, respectively.

Zero or Null Matrix

A matrix is said to be zero matrix or null matrix if all its elements are zero.

For Example,

$$A = [0] \begin{bmatrix} 0 & 0 \\ 0 & 0 \end{bmatrix} \begin{bmatrix} 0 & 0 & 0 \\ 0 & 0 & 0 \\ 0 & 0 & 0 \end{bmatrix}$$

are all zero matrices of the order 1, 2 and 3 respectively. We denote zero matrix by O.

Unit or Identity Matrix

If a square matrix has all elements 0 and each diagonal elements are non-zero, it is called identity matrix and denoted by I.

Equal Matrices: Two matrices are said to be equal if they are of the same order and if their corresponding elements are equal to the square matrix $A = [a_{ij}]_{n \times n}$ is an identity matrix if:

- $a_{ij} = 1$ if $i = j$
- $a_{ij} = 0$ if $i \neq j$

We denote the identity matrix of order n by I_n. When the order is clear from the context, we simply write it as I. For example,

$$A = [1] \begin{bmatrix} 1 & 0 \\ 0 & 1 \end{bmatrix} \begin{bmatrix} 1 & 0 & 0 \\ 0 & 1 & 0 \\ 0 & 0 & 1 \end{bmatrix}$$

are identity matrices of order 1, 2 and 3, respectively. Observe that a scalar matrix is an identity matrix when k = 1. But every identity matrix is clearly a scalar matrix.

Upper Triangular Matrix

A square matrix in which all the elements below the diagonal are zero is known as the upper triangular matrix. For example,

$$A = \begin{bmatrix} 3 & -5 & 7 \\ 0 & 0 & 0 \\ 0 & 0 & 9 \end{bmatrix}$$

Lower Triangular Matrix

A square matrix in which all the elements above the diagonal are zero is known as the upper triangular matrix. For example,

$$A = \begin{bmatrix} 3 & 0 & 0 \\ 0 & 4 & 0 \\ -5 & 7 & 9 \end{bmatrix}$$

Determinant of a Matrix

To every square matrix A = [aij] of order n, we can associate a number (real or complex) called determinant of the square matrix A, where a = (i, j)th element of A. This may be thought of as a function which associates each square matrix with a unique number (real or complex).

If M is the set of square matrices, K is the set of numbers (real or complex) and $f : M \rightarrow K$ is defined by f (A) = k, where $A \in M$ and $k \in K$, then f (A) is called the determinant of A. It is also denoted by | A | or det A or Δ.

If $A = \begin{bmatrix} a & b \\ c & d \end{bmatrix}$, then determinant of A is written as

$$|A| = \begin{vmatrix} a & b \\ c & d \end{vmatrix} = \det A$$

For a 1×1 Matrix

Let A = [a] be the matrix of order 1, then determinant of A is defined to be equal to a.

For a 2 × 2 Matrix

For a 2 × 2 matrix (2 rows and 2 columns):

$$A = \begin{bmatrix} a & b \\ c & d \end{bmatrix}$$

The determinant is: |A| = ad − bc or the determinant of A equals a × d minus b × c. It is easy to remember when you think of a cross, where blue is positive that goes diagonally from left to right and red is negative that goes diagonally from right to left.

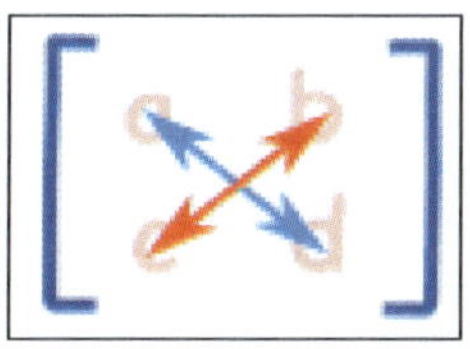

Example: If,

$$A = \begin{bmatrix} 2 & 3 \\ 4 & 8 \end{bmatrix}$$

$$|A| = 2 \times 8 - 4 \times 3$$

$$= 16 - 12$$

$$= 4$$

For a 3 × 3 Matrix

For a 3 × 3 matrix (3 rows and 3 columns):

$$A = \begin{bmatrix} a & b & c \\ d & e & f \\ g & h & i \end{bmatrix}$$

The determinant is: |A| = a (ei – fh) – b (di – fg) + c (dh – eg). The determinant of A equals 'a times e x i minus f x h minus b times d x i minus f x g plus c times d x h minus e x g'. It may look complicated, but if you carefully observe the pattern its really easy.

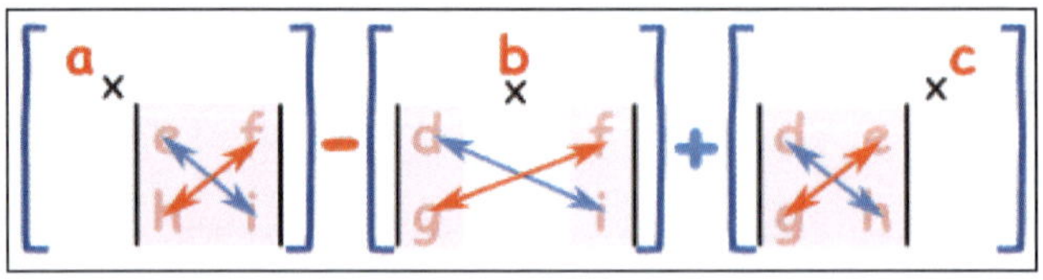

To work out the determinant of a matrix 3×3:

- Multiply 'a' by the determinant of the 2×2 matrix that is not in a's row or column.
- Likewise for 'b' and for 'c'.
- Sum them up, but remember the minus in front of the b.

As a formula *(remember the vertical bars || mean "determinant of")*:

$$|A| = a \cdot \begin{vmatrix} e & f \\ h & i \end{vmatrix} - b \cdot \begin{vmatrix} d & f \\ g & i \end{vmatrix} + c \cdot \begin{vmatrix} d & e \\ g & h \end{vmatrix}$$

The determinant of A equals 'a' times the determinant of e × i minus f × h minus 'b' times the determinant of d × i minus f × g plus 'c' times the determinant of d × h minus e × g.

Example: If,

$$A = \begin{bmatrix} 6 & 1 & 1 \\ 4 & -2 & 5 \\ 2 & 8 & 7 \end{bmatrix}$$

$$|A| = 6\times(-2\times7 - 5\times8) - 1\times(4\times7 - 5\times2) + 1\times(4\times8 - (-2\times2))$$

$$= 6\times(-54) - 1\times(18) + 1\times(36)$$

$$= -306$$

For 4 × 4 Matrices and Higher

The pattern continues for the determinant of a matrix 4×4:

- Plus a times the determinant of the matrix that is not in a's row or column,

- Minus b times the determinant of the matrix that is not in b's row or column,
- Plus c times the determinant of the matrix that is not in c's row or column,
- Minus d times the determinant of the matrix that is not in d's row or column.

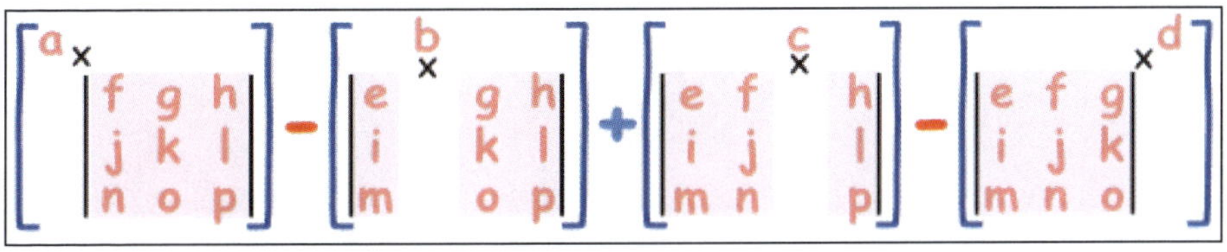

As a formula:

$$|A| = a \cdot \begin{vmatrix} f & g & h \\ j & k & l \\ n & o & p \end{vmatrix} - b \cdot \begin{vmatrix} e & g & h \\ i & k & l \\ m & o & p \end{vmatrix} + c \cdot \begin{vmatrix} e & f & h \\ i & j & l \\ m & n & p \end{vmatrix} - d \cdot \begin{vmatrix} e & f & g \\ i & j & k \\ m & n & o \end{vmatrix}$$

Notice the +−+− pattern (+a... −b... +c... −d...).

Example: If,

$$f(x)\begin{vmatrix} x & \lambda \\ 2\lambda & x \end{vmatrix},$$

Then, $f(\lambda x) - f(x)$ is equal to:

a. $x(\lambda^2 - 1)$

b. $2\lambda(x^2 - 1)$

c. $\lambda^2(x^2 - 1)$

d. $x^2(\lambda^2 - 1)$

Solution: Given,

$$f(x) = \begin{vmatrix} x & \lambda \\ 2\lambda & x \end{vmatrix}$$

$\Rightarrow f(x) = x^2 - 2\lambda^2$

Therefore,

$f(\lambda x) - f(x) = (\lambda x)^2 - 2\lambda^2 - (x^2 - 2\lambda^2)$

$= \lambda^2 x^2 - 2\lambda^2 - x^2 + 2\lambda^2$

$= x^2(\lambda^2 - 1)$

So, option D is correct.

Example: If,

$$A = \begin{vmatrix} 0 & c & -b \\ -c & 0 & a \\ b & -a & 0 \end{vmatrix}$$

then $(a^2 + b^2 - c^2)\ |A| =$

a. abc

b. a + b + c

c. a3 + b3 + c3

d. 0

Solution: $|A| = 0 \times (a^2) - c(-ab) - b(ac) = 0 + abc - abc = 0$.

Transpose of a Matrix

The new matrix obtained by interchanging the rows and columns of the original matrix is called as the transpose of the matrix. If $A = [a_{ij}]$ be an m × n matrix, then the matrix obtained by interchanging the rows and columns of A would be the transpose of A. of It is denoted by A′or (A^T). In other words, if $A = [a_{ij}]_{mxn}$,then $A' = [a_{ji}]_{nxm}$. For example,

$$A = \begin{bmatrix} 3 & -5 \\ 4 & 7/2 \\ 9 & 5/8 \end{bmatrix} \quad A' = \begin{bmatrix} 3 & 4 & 9 \\ -5 & 7/2 & 5/8 \end{bmatrix}$$

Properties

1. Transpose of Transpose of a Matrix

The transpose of the transpose of a matrix is the matrix itself: $(A^T)^T = A$. For example,

$$A = \begin{bmatrix} 3 & 2 & 1 \\ 1 & 3 & 2 \end{bmatrix}$$

Verify that $(A^T)^T = A$. It is determined as shown below:

$$A^T = \begin{bmatrix} 3 & 1 \\ 2 & 3 \\ 1 & 2 \end{bmatrix}$$

Therefore,

$$(\quad) \quad \begin{bmatrix} 3 & 2 & 1 \\ 1 & 3 & 2 \end{bmatrix}$$

2. Transpose of a Scalar Multiple

The transpose of a matrix times a scalar (k) is equal to the constant times the transpose of the matrix: $(kA)^T = kA^T$. For example,

$$Let \ A = \begin{bmatrix} 1 & 2 \\ 3 & 4 \end{bmatrix}$$

and k = 2. Verify that $(kA)^T = kA^T$

Solution: LHS = $(kA)^T$

$$= \left(2 \times \begin{bmatrix} 1 & 2 \\ 3 & 4 \end{bmatrix} \right)^T$$

$$= \left(\begin{bmatrix} 2 & 4 \\ 6 & 8 \end{bmatrix} \right)^T$$

RHS = kA^T

$$= 2 \times \begin{bmatrix} 1 & 2 \\ 3 & 4 \end{bmatrix}^T$$

$$= 2 \times \begin{bmatrix} 1 & 3 \\ 2 & 4 \end{bmatrix}$$

$$= \begin{bmatrix} 2 & 6 \\ 4 & 8 \end{bmatrix}$$

Therefore, LHS = RHS. Hence, $(kA)^T = kA^T$.

3. Transpose of a Sum

The transpose of the sum of two matrices is equivalent to the sum of their transposes: $(A + B)^T = A^T + B^T$. For example:

$$A = \begin{bmatrix} 2 & 3 & 1 \\ 1 & -2 & 7 \end{bmatrix}$$

$$B = \begin{bmatrix} 1 & 2 & 3 \\ 0 & 4 & 6 \end{bmatrix}$$

verify that $(A \pm B)^T = A^T \pm B^T$.

$$A + B = \begin{bmatrix} 3 & 5 & 4 \\ 1 & 2 & 13 \end{bmatrix}$$

Therefore,

$$(A+B)^T = \begin{bmatrix} 3 & 1 \\ 5 & 2 \\ 4 & 13 \end{bmatrix}$$

The transpose matrices for A and B are given as below:

$$A^T = \begin{bmatrix} 2 & 1 \\ 3 & -2 \\ 1 & 7 \end{bmatrix} \text{ and } B^T = \begin{bmatrix} 1 & 0 \\ 2 & 4 \\ 3 & 6 \end{bmatrix}$$

Therefore,

$$A^T \pm B^T = \begin{bmatrix} 3 & 1 \\ 5 & 2 \\ 4 & 13 \end{bmatrix}$$

Hence $(A \pm B)^T = A^T \pm B^T$.

4. Transpose of a Product

The transpose of the product of two matrices is equivalent to the product of their transposes in reversed order: $(AB)^T = B^T A^T$. The same is true for the product of multiple matrices: $(ABC)^T = C^T B^T A^T$. For example,

$$A = \begin{bmatrix} 1 & 2 \\ 3 & 4 \end{bmatrix}$$

$$B = \begin{bmatrix} 1 & 0 \\ 2 & -1 \end{bmatrix}$$

verify that $(AB)^T = B^T A^T$.

Solution: The product of A and B is:

$$(AB) = \begin{bmatrix} 1 & 2 \\ 3 & 4 \end{bmatrix} \begin{bmatrix} 1 & 0 \\ 2 & -1 \end{bmatrix} = \begin{bmatrix} 5 & -2 \\ 11 & -4 \end{bmatrix}$$

Therefore,

$$(AB)^T = \begin{bmatrix} 5 & 11 \\ -2 & -4 \end{bmatrix}$$

If we take the transpose of A and B separately and multiply A with B, then we have:

$$B^T A^T = \begin{bmatrix} 1 & 2 \\ 0 & -1 \end{bmatrix} \begin{bmatrix} 1 & 3 \\ 2 & 4 \end{bmatrix} = \begin{bmatrix} 5 & 11 \\ -2 & -4 \end{bmatrix}$$

Hence $(AB)^T = B^T A^T$.

Example:

If matrix A is a circulant matrix whose elements of first row are a, b, c all > 0 such that abc = 1 and $A^TA = I$, then $a^3 + b^3 + c^3$ equals:

a. 0

b. 3

c. 1

d. 4

Solution: Given, A is a circulant matrix of elements a, b, c and abc = 1. So,

$$A = \begin{bmatrix} a & b & c \\ c & a & b \\ b & c & a \end{bmatrix}$$

So, det A = $a(a^2 - bc) - b(ac-b^2) + c(c^2 - ab) = a^3 + b^3 + c^3 - 3$ and $A^TA = I$

$|A^TA| = |I| = 1$

$|A|\,|A^T| = 1$

$|A|^2 = 1$

Therefore, $|A| = \pm 1$. So, det A = $\pm$ 1. After substituting the value in the det A = $a^3 + b^3 + c^3 - 3abc$, we get, $a^3 + b^3 + c^3 = 4$ or $a^3 + b^3 + c^3 = -2$. Therefore, answer is option D.

Inverse of a Matrix

We know that the multiplicative inverse of a real number a is a^{-1}, and $aa^{-1} = a^{-1}a = \left(\frac{1}{a}\right)a = 1$. The multiplicative inverse of a matrix is similar in concept, except that the product of matrix A and its inverse A^{-1} equals the identity matrix. The identity matrix is a square matrix containing ones down the main diagonal and zeros everywhere else. We identify identity matrices by I_n where n represents the dimension of the matrix. The next two equations are the identity matrices for a 2 × 2 matrix and a 3 ×3 matrix, respectively.

$$I_3 = \begin{bmatrix} 1 & 0 & 0 \\ 0 & 1 & 0 \\ 0 & 0 & 1 \end{bmatrix}$$

The identity matrix acts as a 1 in matrix algebra. For example,

$$AI = IA = A$$

A matrix that has a multiplicative inverse has the properties,

$$\begin{bmatrix} AA^{-1} = I \\ A^{-1}A = I \end{bmatrix}$$

A matrix that has a multiplicative inverse is called an invertible matrix. Only a square matrix may have a multiplicative inverse, as the reversibility, $AA^{-1} = A^{-1}A = I$, is a requirement. Not all square matrices have an inverse, but if $A$is invertible, then A^{-1} is unique. We will look at two methods for finding the inverse of a 2 × 2 matrix and a third method that can be used on both 2 × 2and 3 × 3 matrices.

Methods to Find the Inverse of a Matrix

Method 1: Transposing and Determinants

This method is only good for finding the inverse of a 2 × 2 matrix.

Example: Find the inverse, A^{-1}, of:

$$A = \begin{pmatrix} 2 & -3 \\ 4 & -7 \end{pmatrix}$$

using Method 1.

Method 1 is as follows.

1. Interchange leading diagonal elements:

   $-7 \to 2; \quad 2 \to -7$

   $$\begin{pmatrix} -7 & -3 \\ 4 & 2 \end{pmatrix}$$

2. Change signs of the other 2 elements:

   $-3 \to 3; \quad 4 \to -4$

   $$\begin{pmatrix} -7 & 3 \\ -4 & 2 \end{pmatrix}$$

3. Find the determinant |A|

$$\begin{vmatrix} 2 & -3 \\ 4 & -7 \end{vmatrix} = -14 + 12 = -2$$

4. Multiply result of by $\frac{1}{|A|}$

$$A^{-1} = \frac{1}{|A|}\begin{pmatrix} -7 & 3 \\ -4 & 2 \end{pmatrix}$$

$$= \frac{1}{-2}\begin{pmatrix} -7 & 3 \\ -4 & 2 \end{pmatrix}$$

$$= \begin{pmatrix} 3.5 & -1.5 \\ 2 & -1 \end{pmatrix}$$

So, we have found the inverse, as required.

Method 2: Adjunct Matrix (can be Extended to Any Size)

Method 2 uses the adjoint matrix method.

The inverse of a 3×3 matrix is given by:

$$A^{-1} = \frac{\text{adj A}}{\text{det A}}$$

"adj A" is short for "the adjoint of A". We use cofactors to determine the adjoint of a matrix.

Cofactors

The cofactor of an element in a matrix is the value obtained by evaluating the determinant formed by the elements not in that particular row or column.

Examples:

1. Consider the matrix:

$$\begin{pmatrix} 5 & 6 & 1 \\ 0 & 3 & -3 \\ 4 & -7 & 2 \end{pmatrix}$$

The cofactor of 6 is:

$$\begin{vmatrix} 0 & -3 \\ 4 & 2 \end{vmatrix} = 0 + 12 = 12$$

The cofactor of -3 is:

$$\begin{vmatrix} 5 & 6 \\ 4 & -7 \end{vmatrix} = -35 - 24 = -59$$

We find the adjoint matrix by replacing each element in the matrix with its cofactor and applying a + or - sign as follows:

$$\begin{pmatrix} + & - & + \\ - & + & - \\ + & - & + \end{pmatrix}$$

and then finding the transpose of the resulting matrix. The transpose means the 1st column becomes the 1st row; 2nd column becomes 2nd row, etc.

2. Find the inverse of the following by using the adjoint matrix method:

$$A = \begin{pmatrix} 5 & 6 & 1 \\ 0 & 3 & -3 \\ 4 & -7 & 2 \end{pmatrix}$$

Solution:

Step 1: Replace elements with cofactors and apply + and –

$$\begin{pmatrix} +(-15) & -(12) & +(-12) \\ -(19) & +(6) & -(-59) \\ +(-21) & -(-15) & +(15) \end{pmatrix}$$

$$= \begin{pmatrix} -15 & -12 & -12 \\ -19 & 6 & 59 \\ -21 & 15 & 15 \end{pmatrix}$$

Step 2: Transpose the matrix:

$$adj\ A = \begin{pmatrix} -15 & -19 & -21 \\ -12 & 6 & 15 \\ -12 & 59 & 15 \end{pmatrix}$$

Before we can find the inverse of matrix A, we need det A:

$$\begin{vmatrix} 5 & 6 & 1 \\ 0 & 3 & -3 \\ 4 & -7 & 2 \end{vmatrix} = 5(-15) + 4(-21) = -159$$

Now we have what we need to apply the formula:

$$A^{-1} = \frac{\text{adj A}}{\text{det A}}$$

So,

$$A^{-1} = \frac{\text{adj A}}{\text{det A}}$$

$$= \frac{1}{-159}\begin{pmatrix} -15 & -19 & -21 \\ -12 & 6 & 15 \\ -12 & 59 & 15 \end{pmatrix}$$

$$= \begin{pmatrix} 0.094 & 0.119 & 0.132 \\ 0.075 & -0.038 & -0.094 \\ 0.075 & -0.371 & -0.094 \end{pmatrix}$$

3. Find the inverse of:

$$\begin{pmatrix} -2 & 6 & 1 \\ 0 & 3 & -3 \\ 4 & -7 & 3 \end{pmatrix}$$

using Method 2.

Solution:

$$C\ of\ A = \begin{pmatrix} +(-12) & -(12) & +(-12) \\ -(25) & +(-10) & -(-10) \\ +(-21) & -(6) & +(-6) \end{pmatrix}$$

$$= \begin{pmatrix} -12 & -12 & -12 \\ -25 & -10 & 10 \\ -21 & -6 & -6 \end{pmatrix}$$

Interchange rows and columns:

$$adj\ A = \begin{pmatrix} -12 & -25 & -21 \\ -12 & -10 & -6 \\ -12 & 10 & -6 \end{pmatrix}$$

$$\begin{aligned} det\ A &= \begin{vmatrix} -2 & 6 & 1 \\ 0 & 3 & -3 \\ 4 & -7 & 3 \end{vmatrix} \\ &= 2(9-21)+4(-21) \\ &= -60 \end{aligned}$$

So,

$$\begin{aligned} A^{-1} &= \frac{\text{adj A}}{\text{det A}} \\ &= \frac{1}{-60}\begin{pmatrix} -12 & -25 & -21 \\ -12 & -10 & -6 \\ -12 & 10 & -6 \end{pmatrix} \\ &= \begin{pmatrix} \frac{1}{5} & \frac{5}{12} & \frac{7}{20} \\ \frac{1}{5} & \frac{1}{6} & \frac{1}{10} \\ \frac{1}{5} & -\frac{1}{6} & \frac{1}{10} \end{pmatrix} \\ &= \begin{pmatrix} 0.2 & 0.417 & 0.35 \\ 0.2 & 0.167 & 0.1 \\ 0.2 & -0.167 & 0.1 \end{pmatrix} \end{aligned}$$

Method 3: Inverses of Larger Matrices

Most real systems of equations are very large (up to 100 by 100 is common). We use computers to find these inverses.

Find the inverse of:

$$\begin{pmatrix} 7 & -2 \\ -6 & 2 \end{pmatrix}$$

by Method 1.

1. Interchange leading diagonal elements:

$$\begin{pmatrix} 2 & -2 \\ -6 & 7 \end{pmatrix}$$

2. Change signs of the other 2 elements:

$$\begin{pmatrix} 2 & 2 \\ 6 & 7 \end{pmatrix}$$

3. Find |A|: Remember that our original matrix (from the question) is:

$$A = \begin{pmatrix} 7 & -2 \\ -6 & 2 \end{pmatrix}$$

So, the determinant of A is given by:

$$|A| = \begin{vmatrix} 7 & -2 \\ -6 & 2 \end{vmatrix} = 14 - 12 = 2$$

4. Multiply result of by $\frac{1}{|A|}$

$$A^{-1} = \frac{1}{|A|}\begin{pmatrix} 2 & 2 \\ 6 & 7 \end{pmatrix}$$
$$= \frac{1}{2}\begin{pmatrix} 2 & 2 \\ 6 & 7 \end{pmatrix}$$
$$= \begin{pmatrix} 1 & 1 \\ 3 & 3.5 \end{pmatrix}$$

Check,

$$A^{-1}A = \begin{pmatrix} 1 & 1 \\ 3 & 3.5 \end{pmatrix}\begin{pmatrix} 7 & -2 \\ -6 & 2 \end{pmatrix}$$
$$= \begin{pmatrix} 7-6 & -2+2 \\ 21-21 & -6+7 \end{pmatrix}$$
$$= \begin{pmatrix} 1 & 0 \\ 0 & 1 \end{pmatrix}$$
$$= I$$

Adjoint of a Matrix

The adjoint of a square matrix A = $[a_{ij}]_{n \times n}$ is defined as the transpose of the matrix $[A_{ij}]_{n \times n}$, where Aij is the cofactor of the element a_{ij}. Adjoing of the matrix A is denoted by adj A.

Thus is,

$$A = \begin{bmatrix} a_{11} & a_{12} & \cdots & a_{1n} \\ a_{21} & a_{22} & \cdots & a_{2n} \\ \cdots & \cdots & \cdots & \cdots \\ a_{n1} & a_{n2} & \cdots & a_{nn} \end{bmatrix}$$

$$\text{Adj. A} = \text{Transpose of} \begin{bmatrix} A_{11} & A_{12} & \cdots & A_{1n} \\ A_{21} & A_{22} & \cdots & A_{2n} \\ A_{n1} & A_{n2} & \cdots & A_{nn} \end{bmatrix}$$

$$= \begin{bmatrix} A_{11} & A_{21} & \cdots & A_{n1} \\ A_{12} & A_{22} & \cdots & A_{n2} \\ \vdots & & & \\ A_{1n} & A_{12} & \cdots & A_{nn} \end{bmatrix}$$

Examples:

1. Find the adjoint of the matrix:

$$\begin{bmatrix} 1 & -1 & 2 \\ 4 & 0 & 6 \\ 0 & 1 & -1 \end{bmatrix}$$

Solution: We will first evaluate the cofactor of every element,

$$cof\left(a_{11}\right) = +\begin{vmatrix} 0 & 6 \\ 1 & -1 \end{vmatrix} = -6 \qquad cof\left(a_{12}\right) = -\begin{vmatrix} 4 & 6 \\ 0 & -1 \end{vmatrix} = 4$$

$$cof\left(a_{21}\right) = -\begin{vmatrix} -1 & 2 \\ 1 & -1 \end{vmatrix} = 1 \qquad cof\left(a_{22}\right) = -\begin{vmatrix} 1 & 2 \\ 0 & -1 \end{vmatrix} = -1$$

$$cof\left(a_{31}\right) = +\begin{vmatrix} -1 & 2 \\ 0 & 6 \end{vmatrix} = -6 \qquad cof\left(a_{32}\right) = -\begin{vmatrix} 1 & 2 \\ 4 & 6 \end{vmatrix} = 2$$

$$cof\left(a_{13}\right) = +\begin{vmatrix} 4 & 0 \\ 0 & 1 \end{vmatrix} = 4$$

$$cof\left(a_{23}\right) = -\begin{vmatrix} 1 & -1 \\ 0 & 1 \end{vmatrix} = -1$$

$$cof\left(a_{33}\right) = +\begin{vmatrix} 1 & -1 \\ 4 & 4 \end{vmatrix} = 4$$

Therefore,

$$Adj\ A = \left[cof\left(a_{ij}\right)\right]^T = \begin{bmatrix} -6 & 4 & 4 \\ 1 & -1 & -1 \\ -6 & 2 & 4 \end{bmatrix}^T = \begin{bmatrix} -6 & 1 & -6 \\ 4 & -1 & 2 \\ 4 & -1 & 4 \end{bmatrix}$$

The Relation between Adjoint and Inverse of a Matrix

To find the inverse of a matrix A, i.e A-1 we shall first define the adjoint of a matrix. Let A be an n x n matrix. The (i,j) cofactor of A is defined to be:

$$A_{ij} = (-1)^{ij} \det\left(M_{ij}\right),$$

where, M_{ij} is the (i,j)th minor matrix obtained from A after removing the ith row and jth column. Let's consider the n x n matrix A = (Aij) and define the n x n matrix Adj(A) = A^T. The matrix Adj(A) is called the adjoint of matrix A. When A is invertible, then its inverse can be obtained by the formula given below.

$$A^{-1} = \frac{adj\ A}{|A|}$$

The inverse is defined only for non-singular square matrices. The following relationship holds between a matrix and its inverse:

$AA^{-1} = A^{-1}A = I$, where I is the identity matrix.

2. Determine whether the matrix given below is invertible and if so, then find the invertible matrix using the above formula.

$$\begin{bmatrix} 1 & 5 & 2 \\ 0 & -1 & 2 \\ 0 & 0 & 1 \end{bmatrix}$$

Solution: Since A is an upper triangular matrix, the determinant of A is the product of its diagonal entries. This, we have det(A) = -1, which is a non-zero value and hence, A is invertible. To find the inverse using the formula, we will first determine the cofactors A_{ij} of A. We have,

$$A_{11} = \begin{vmatrix} -1 & 2 \\ 0 & 1 \end{vmatrix} = -1, \quad A_{12} = -\begin{vmatrix} 0 & 2 \\ 0 & 1 \end{vmatrix} = 0, \quad A_{13} = \begin{vmatrix} 0 & -1 \\ 0 & 0 \end{vmatrix} = 0$$

$$A_{21} = -\begin{vmatrix} 5 & 2 \\ 0 & 1 \end{vmatrix} = -5, \quad A_{22} = \begin{vmatrix} 1 & 2 \\ 0 & 1 \end{vmatrix} = 1, \quad A_{23} = -\begin{vmatrix} 1 & 5 \\ 0 & 0 \end{vmatrix} = 0$$

$$A_{31} = \begin{vmatrix} 5 & 2 \\ -1 & 2 \end{vmatrix} = 12, \quad A_{32} = -\begin{vmatrix} 1 & 2 \\ 0 & 2 \end{vmatrix} = -2, \quad A_{33} = \begin{vmatrix} 1 & 5 \\ 0 & -1 \end{vmatrix} = -1.$$

Then the adjoint matrix of A is:

$$\begin{bmatrix} -1 & -5 & 12 \\ 0 & 1 & -2 \\ 0 & 0 & -1 \end{bmatrix}$$

Using the formula, we will obtain the inverse matrix as:

$$A^{-1} = adj\ (A)/\det(A) = \begin{bmatrix} 1 & 5 & -12 \\ 0 & -1 & 2 \\ 0 & 0 & 1 \end{bmatrix}$$

Theorems on Adjoint and Inverse of a Matrix

1. If A be any given square matrix of order n, then A adj(A) = adj(A) A = |A|I, where I is the identitiy matrix of order n.

   Proof: Let,

   $$A = \begin{bmatrix} a_{11} & a_{12} & a_{13} \\ a_{21} & a_{22} & a_{23} \\ a_{31} & a_{32} & a_{33} \end{bmatrix}, \text{ then } adj\ A = \begin{bmatrix} A_{11} & A_{21} & A_{31} \\ A_{12} & A_{22} & A_{32} \\ A_{13} & A_{23} & A_{33} \end{bmatrix}$$

   Since the sum of the product of elements of a row (or a column) with corresponding cofactors is equal to |A| and otherwise zero, we have:

   $$A(adj\ A) = \begin{bmatrix} |A| & 0 & 0 \\ 0 & |A| & 0 \\ 0 & 0 & |A| \end{bmatrix} = |A| \begin{bmatrix} 1 & 0 & 0 \\ 0 & 1 & 0 \\ 0 & 0 & 1 \end{bmatrix} = |A|\ I$$

   Similarly, we can show that adj(A) A = |A| I

   Hence, A adj(A) = adj(A) A

2. If A and B are non-singular matrices of the same order, then AB and BA are also non-singular matrices of the same order.

3. The determinant of the product matrices is equal to the product of their respective determinants, that is, |AB| = |A||B|, where A and B are square matrices of the same order.

   We know that:

   $$(adj\ A)A = |A|I = \begin{bmatrix} |A| & 0 & 0 \\ 0 & |A| & 0 \\ 0 & 0 & |A| \end{bmatrix}$$

Writing determinants of matrices on both sides, we have:

$$|(adj\ A)A| = \begin{vmatrix} |A| & 0 & 0 \\ 0 & |A| & 0 \\ 0 & 0 & |A| \end{vmatrix}$$

i.e.,

$$|(adj\ A)||A| = |A|^3 \begin{vmatrix} 1 & 0 & 0 \\ 0 & 1 & 0 \\ 0 & 0 & 1 \end{vmatrix}$$

i.e., $|adj(A)|\ |A| = |A|^3$

or $|adj(A)| = |A|^2$

In general, if A is a quare matrix of order n, then $|adj(A)| = |A|^{n-1}$.

4. A square matrix A is invertible if and only if A is a non-singular matrix.

   Proof: Let A be an invertible matrix of order n and I be the identity matrix of the same order. Then there exists a square matrix B of order n such that AB = BA = I. Now, AB = I.

   So, $|A|\ |B| = |I| = 1$ (since $|I| = 1$ and $|AB| = |A|\ |B|$). This gives $|A|$ to be a non-zero value. Hence A is a non-singular matrix. Conversely, let A be a non-singular matrix, then $|A|$ is non-zero. Now A adj(A) = adj(A) A = $|A|$ I (Theorem 1). Or,

$$A\left(\frac{1}{|A|} adj\ A\right) = \left(\frac{1}{|A|} adj\ A\right) A = I$$

   or AB = BA = I, where

$$B = \frac{1}{|A|} adj\ A$$

   Thus, A is invertible and

$$A^{-1} = \frac{1}{|A|} adj\ A$$

Example: With 1, ω, ω2 as cube roots of units, inverse of which of the following matrices exist?

A. $\begin{bmatrix} 1 & \omega \\ \omega & \omega^2 \end{bmatrix}$

B. $\begin{bmatrix} \omega^2 & 1 \\ 1 & \omega \end{bmatrix}$

$$C.\begin{bmatrix} \omega & \omega^2 \\ \omega^2 & 1 \end{bmatrix}$$

D. None of these.

Solution: Option D. Inverse of a matrix exists if its determinant if not equal to 0. For option A, B and C, the determinants are equal to zero, hence the inverse does not exist for any of the given matrices.

Rank of a Matrix

The row rank of a matrix is the maximum number of rows, thought of as vectors, which are linearly independent. Similarly, the column rank is the maximum number of columns which are linearly independent. It is an important result, not too hard to show that the row and column ranks of a matrix are equal to each other. Thus one simply speaks of the rank of a matrix.

We will show this for 3 × 2 matrices – essentially without relying on linear algebra. Let,

$$A = \begin{pmatrix} a_1 & b_1 \\ a_2 & b_2 \\ a_3 & b_3 \end{pmatrix}$$

If the column rank is zero, clearly all entries are zero and the statement is obvious.

If the column rank is one, it means that $a = (a_1, a_2, a_3)$ is a multiple of $b = (b_1, b_2, b_3)$ (or vice-versa), or $a_1 = \lambda b_1$, $a_2 = \lambda b_2$ and $a_3 = \lambda b_3$ for some non-zero λ and;

$$A = \begin{pmatrix} \lambda b_1 & b_1 \\ \lambda b_2 & b_2 \\ \lambda b_3 & b_3 \end{pmatrix}$$

Now, the row vectors are: $b_1 (\lambda, 1)$, $b_2 (\lambda, 1)$ and $b_3 (\lambda, 1)$, all multiple of the same nonzero vector $(\lambda, 1)$, so there is one and only one linearly independent row.

Finally, assume that the column rank is 2, meaning a and b are linearly independent. Then both are nonzero and the angle θ between them is not 0 or π, or equivalently, $\sin\theta \neq 0$ or, equivalently still, $a \times b \neq 0$. But by definition,

$$a \times b = \begin{pmatrix} i & j & k \\ a_1 & a_2 & a_3 \\ b_1 & b_2 & b_3 \end{pmatrix} = i\begin{vmatrix} a_2 & a_3 \\ b_2 & b_3 \end{vmatrix} - j\begin{vmatrix} a_1 & a_3 \\ b_1 & b_3 \end{vmatrix} + k\begin{vmatrix} a_1 & a_2 \\ b_1 & b_2 \end{vmatrix}$$

where, | | denotes as usual the determinant.

For a × b to be nonzero, at least one of the determinants above has to be nonzero. Say the last one is nonzero. But then we cannot have $a_1 = \lambda a_2$ and $b_1 = \lambda b_2$ since otherwise the determinant would be $a_1b_2 - a_2b_1 = \lambda a_2b_2 - \lambda b_2a_2 = 0$. Thus (a_1, b_1) is linearly independent from (a_2, b_2). Of course no more than two vectors with 2 components (that is, two vectors in $\mathbb{R}^2$ can be linearly independent, and thus the row rank is also 2.

Vector Space

Vector spaces are mathematical objects that abstractly capture the geometry and algebra of linear equations. They are the central objects of study in linear algebra.

The archetypical example of a vector space is the Euclidean space $\mathbb{R}^n$. In this space, *vectors* are n-tuples of real numbers; for example, a vector in $\mathbb{R}^2$ (3, 4). These vectors have an addition operation defined on them, where one adds coordinate-wise:

$$(a_1, a_2, ...a_n) + (b_1, b_2, ...b_n) = (a_1 + b_1, a_2 + b_2, ...a_n + b_n).$$

For a numerical example, consider the equation (1,-1) + (6, 1) = (7,0), which holds in $\mathbb{R}^2$. There is also a notion of *scalar multiplication*; given a number $c \in \mathbb{R}$, one may scale a vector by cc as follows:

$$c \cdot (a_1, a_2, ..., a_n) = (ca_1, ca_2, ..., ca_n).$$

For instance, in $\mathbb{R}^2$ one has 3 · (5, 8) = (15, 24).

Abstract vector spaces generalize the example of $\mathbb{R}^n$. Roughly, a vector space is a set whose elements are called *vectors*, and these vectors can be added and scaled according to a set of axioms modelled on properties of $\mathbb{R}^n$.

Vector spaces often arise as solution sets to various problems involving linearity, such as the set of solutions to homogeneous system of linear equations and the set of solutions of a homogeneous linear differential equation.

Apart from their central role in linear algebra, vector spaces, equipped with some additional structure, appear frequently in other areas of mathematics. Prominent examples include Hilbert spaces and the use of vector spaces in representation theory.

Let F be a field; for simplicity, one can assume $F = \mathbb{R}$ or $F = C$.

A vector space over FF is an abelian group (V, +) such that for every $c \in F$, there is a map $\phi_c : V \rightarrow V$ satisfying the following properties:

- For any v, w ∈ V and c ∈ F, we have:

  $$\phi_c(v + w) = \phi_c(v) + \phi_c(w).$$

  In other words, the maps ϕ_c are group homomorphisms of V.

- The map $\phi_1 : V \to V$ is just the identity map on V. That is, for any $v \in V$, one has $\phi_1(v) = v$.
- Composition of the maps ϕ_c corresponds to multiplication in F : if a, b $a, b \in F$, then:

  $\phi_a \circ \phi_b = \phi_{ab}$.
- Addition in F commutes with addition in V : if a, b ∈ F and ∈ V, then:

  $\phi_{a+b}(v) = \phi_a(v) + \phi_b(v)$.

Usually, one simply writes cv to denote $\phi_c(v)$. The map ϕ_c should be thought of as *scaling* the vector v by a factor of c.

To better understand this definition, some examples are in order:

Examples:

1. Verify that $\mathbb{R}^2$ is a vector space over $\mathbb{R}$ under the standard notions of vector addition and scalar multiplication.

   First, we must show $\mathbb{R}^2$ is an abelian group under addition of vectors. Certainly the addition operation is associative and commutative, since addition in $\mathbb{R}$ is associative and commutative. There is an additive identity, the zero vectors (0, 0). For any (a, b) $\in \mathbb{R}^2$, an additive inverse exists, namely $(-a, -b)$. Thus, ($\mathbb{R}^2$, +) is an abelian group.

   Next, we check that scalar multiplication c (a, b) = (ca, cb) satisfies the given properties. We have 1(a, b) = (a, b), as desired. Furthermore,

$$\begin{aligned} c((a,b)+(x,y)) &= c(a+x,\ b+y) \\ &= (ca+cx,\ cb+cy) \\ &= (ca,cb)+(cx,cy) \\ &= c(a,b)+c(x,y) \end{aligned}$$

$$\begin{aligned} mn(a,b) &= (mna,mnb) \\ &= m(na,nb) \\ &= m(n,(a,b)) \end{aligned}$$

$$\begin{aligned} (c+d)(a,b) &= (ca+da,cb+db) \\ &= (ca,cb)+(da,db) \\ &= c(a,b)+d(a,b). \end{aligned}$$

We conclude $\mathbb{R}^2$, with the given addition and scalar multiplication operations, forms a vector space.

2. Other examples of vector spaces include the following:

- For any field F, the set of n-tuples of elements in FF, denoted F^n, is a vector space over F. Vector addition and scalar multiplication are defined precisely as in the case of $\mathbb{R}^n$ above. Taking $F = Z^2$, linear binary codes arise as vector subspaces of $\mathbb{Z}_2^n$.
- Let V = C[0, 1], the set of continuous functions $[0,1] \to \mathbb{R}$. This is a vector space over $\mathbb{R}$. Vector addition is defined by:

  $$(f+g)(x) = f(x) + g(x)$$

  and scalar multiplication by:

  $$(cf)(x) = cf(x).$$

Similarly, $C^k(R)$, the space of k-times continuously differentiable functions $\mathbb{R} \to \mathbb{R}$ is a vector space over $\mathbb{R}$.

- The set $P_n = \left\{\sum_{i=0}^{n} a_i x^i \mid a_i \in \mathbb{R}\, \forall i\right\}$ of polynomials with real coefficients of degree $\leq n$, with the usual polynomial addition and multiplication by a real number, is a vector space.
- $\mathbb{C}$ is a vector space over $\mathbb{R}$.
- $\mathbb{R}$ is a vector space over $\mathbb{Q}$.
- $M_n(\mathbb{R})$, the space of matrices over $\mathbb{R}$ of order n with the matrix addition and scalar multiplication by $c \in R$ defined by multiplying each entry of a matrix by c.

Main Concepts

A set $S = \{v_1,...,v_k\}$ of vectors in a vector space V over F is said to be linearly independent if no vector in S can be expressed as a linear combination of other vectors *S*. One can define linear independence in an equivalent way:

A set $V\{v_1,...,v_k\} \subseteq V$ is linearly independent if $a_1 v_1 + a_2 v_2 + \cdots + a_k v_k = 0$ with $a_i \in F \,\forall\, / = 1,...,k$ implies $a_1 = a_2 = \cdots = a_k = 0$.

The linear span of a subset $S \subseteq V$ is the set of all linear combinations of vectors in S.

A basis of a vector space V is a linearly independent set whose linear span equals V. One of the theorems equivalents to the axiom of choice is that every vector space has a basis.

Having defined a mathematical object, it is natural to consider transformations which preserve its underlying structure. This gives rise to the concept of linear transformation between vector spaces over the same field. The choice of bases of vector spaces V, W over the same field F enables to represent linear transformations between V and W through matrices.

Examples:

1. Vectors in $\mathbb{R}^2$

The set $S_1 = \{(1, 0), (2, 0)\}$ of vectors in $\mathbb{R}^2$ is not linearly independent: $(2, 0) = 2 \cdot (1, 0)$. Geometrically, the span of S is the *xx*-axis in $\mathbb{R}^2$.

The set $S_2 = \{(1, 0), (0, 1)\}$ is linearly independent and its span is $\mathbb{R}^2$. S_2 is a basis of $\mathbb{R}^2$.

The set $S_3 = \{(1, 0), (0, 1), (1, 1)\}$. Although the span of $\mathbb{R}^2$ S_3 is not a basis of $\mathbb{R}^2$ since it is not linearly independent.

2. The set $\{1, x, x_2,..., x^n\}$ is a basis of P_n.

Subspaces

Given a vector space V, it is natural to consider properties of its subspaces. The following theorem provides a useful criterion to find subspaces which are vector spaces with the structure inherited from V:

Let $W \neq \theta$ be a subset of a vector space V over F with the vector space operations from V restricted to W. Then W is a vector space if and only if the following properties hold:

$u + v \in W$ for all $u, v \in W$

$cu \in W$ for all $\in W, c \in F$.

Examples:

1. The set of solutions (x, y, z) $\in \mathbb{R}^3$ of the equation ax + by + cz + d = 0 with a, b, c, d $\in \mathbb{R}$ is a plane in $\mathbb{R}^3$. When d = 0, the plane passes through the origin and forms a vector space as a subspace of $\mathbb{R}^3$.

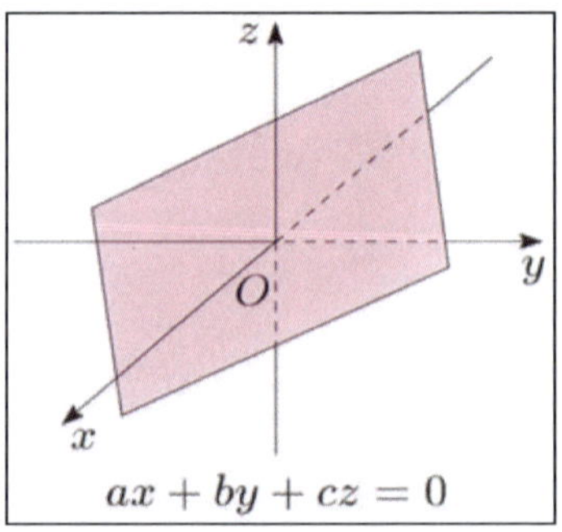

However, planes which do not pass through the origin do not have the structure of a vector space when considered as subsets of points of $\mathbb{R}^3$. This shows the necessity of the restriction to homogeneous systems of linear equations. This technical difficulty is resolved using the concept of affine spaces.

2. The set of functions in $C^2(\mathbb{R})$ satisfying the differential equation $f''(x) = e^x f(x)$ is a vector space.

Since the set of solutions is a nonempty - f(x) = 0 solves the equation - subset of the vector space $C^2(\mathbb{R})$, it suffices to show that if f_1 and f_2 satisfy the given equation, then so do $f_1 + f_2$ and cf_1 for all $c \in \mathbb{R}$. If f_1, f_2 satisfy $f''(x) = e^x f(x)$, then:

$$(f_1(x) + f_2(x))'' = f_1''(x) + f_2''(x) = e^x f_1(x) + e^x f_2(x) + e^x (f_1(x) + f_2(x)).$$

Similarly,

$$\left(cf_1(x)\right)'' = cf_1''(x) = ce^z f_1(x).$$

Additional Structure on Vector Spaces

Examples:

1. An n × n matrix with entries in $\mathbb{R}$ with the property that the sum of entries along each row, column and diagonal is constant and equals a ∈ $\mathbb{R}$ is called a matrix magic square of order n with line-sum a.

   An example of a matrix magic square of order 3 is the matrix:

   $$\begin{pmatrix} 8 & 1 & 6 \\ 3 & 5 & 7 \\ 4 & 9 & 2 \end{pmatrix}$$

   which is itself a magic square. Note that for matrix magic squares we do not have a restriction that the entries of the matrix have to be distinct, so the matrix:

   $$\begin{pmatrix} 0 & 0 & 0 \\ 0 & 0 & 0 \\ 0 & 0 & 0 \end{pmatrix}$$

   is also a matrix magic square of order 3.

The space M of matrix magic squares of order n having arbitrary line-sums forms a vector space as a subspace of the matrix vector space $M_n(\mathbb{R})$:

- The zero matrix is an element of M.
- If A, B ∈ M with line-sums a and b, then the matrix A + B is a matrix magic square with the line-sum a + b. Similarly, the matrix c A is a matrix magic square having the line-sum ca.

Subspace

A subspace is a vector space that is entirely contained within another vector space. As a subspace is defined *relative* to its containing space, both are necessary to fully define one; for example, $\mathbb{R}^2$ is a subspace of $\mathbb{R}^3$, but also of $\mathbb{R}^4$, $\mathbb{C}^2$, etc.

The concept of a subspace is prevalent throughout abstract algebra; for instance, many of the common examples of a vector space are constructed as subspaces of $\mathbb{R}^n$. Subspaces are also useful in analyzing properties of linear transformations, as in the study of fundamental subspaces and the fundamental theorem of linear algebra.

Let V be a vector space. W*W* is said to be a subspace of V if W is a subset of V and the following hold:

- If $w_1, w_2 \in W$, then $w_1 + w_2 \in W$.

- For any scalar c (e.g. a real number), if $w \in W$ then $cw \in W$.

It can be shown that these two conditions are sufficient to ensure W is itself a vector space, as it inherits much of the structure present in V and thus satisfies the remaining conditions on a vector space.

All vector spaces have at least two subspaces: the subspace consisting entirely of the 0 vector, and the "subspace" V itself. These are called the trivial subspaces and rarely have independent significance.

Examples:

The simplest way to generate a subspace is to restrict a given vector space by some rule. For instance, consider the set W of complex vectors v such that:

$$\begin{pmatrix} 2 \\ -3 \\ 4 \end{pmatrix} \cdot v = 0$$

Of course, the set of v satisfying this property is a subset of $\mathbb{C}^3$, by the definition of "complex vector". Thus, if it can be shown that this is in fact a vector space, a subspace of $\mathbb{C}^3$ would be established.

If $w_1, w_2 \in W$, then:

$$\begin{pmatrix} 2 \\ -3 \\ 4 \end{pmatrix} \cdot w_1 = 0, \begin{pmatrix} 2 \\ -3 \\ 4 \end{pmatrix} \cdot w_2 = 0$$

by definition. Hence,

$$\begin{pmatrix} 2 \\ -3 \\ 4 \end{pmatrix} \cdot (w_1 + w_2) = 0$$

and so, $w_1 + w_2 \in W$.

Similarly, if $w \in W$ and $\in C$, then:

$$\begin{pmatrix} 2 \\ -3 \\ 4 \end{pmatrix} \cdot w = 0 \Rightarrow \begin{pmatrix} 2 \\ -3 \\ 4 \end{pmatrix} \cdot (cw) = 0$$

So, $cw \in W$ as desired.

Thus, W is a subspace of V.

Not all subsets form subspaces however. For example, consider the set W of complex vectors $v \in \mathbb{C}^3$ satisfying:

$$\begin{pmatrix} 2 \\ -3 \\ 4 \end{pmatrix} v = 12$$

This is not a subspace as any subspace must inherit the zero vector of its parent, but $v = \begin{pmatrix} 0 \\ 0 \\ 0 \end{pmatrix} \notin W$.

This can also be seen from the above conditions:

$$\begin{pmatrix} 2 \\ -3 \\ 4 \end{pmatrix} \cdot \begin{pmatrix} 6 \\ 0 \\ 0 \end{pmatrix} = 12 \Rightarrow \begin{pmatrix} 6 \\ 0 \\ 0 \end{pmatrix} \in W$$

$$\begin{pmatrix} 2 \\ -3 \\ 4 \end{pmatrix} \cdot \begin{pmatrix} 0 \\ 0 \\ 3 \end{pmatrix} = 12 \Rightarrow \begin{pmatrix} 0 \\ 0 \\ 3 \end{pmatrix} \in W$$

But,

$$\begin{pmatrix} 6 \\ 0 \\ 0 \end{pmatrix} + \begin{pmatrix} 0 \\ 0 \\ 3 \end{pmatrix} = \begin{pmatrix} 6 \\ 0 \\ 3 \end{pmatrix} \text{ is not in W.}$$

Fundamental Subspaces

The fundamental subspaces are four vector spaces defined by a given $m \times n$ matrix A (and its transpose): the column space and nullspace (or kernel) of A, the column space of A^T ((also called the *row space* of, A), and the nullspace of A^T (also called the *left nullspace* of A)).

The fundamental subspaces are useful for a number of linear algebra applications, including analyzing the rank of a matrix. The subspaces are also closely related by the fundamental theorem of linear algebra.

Column Space

The column space of a matrix A is the vector space formed by the columns of A, essentially meaning all linear combinations of the columns of A. Equivalently, the column space consists of all matrices Ax for some vector x.

For this reason, the column space is also known as the image of A (denoted im(A)), as it is the result when A is viewed as a linear transformation of the vector space $\mathbb{R}^m$ (where m is the number of rows of A). In particular, the image of A is necessarily a subspace of $\mathbb{R}^m$, hence the term "fundamental subspace."

For example, consider the matrix:

$$A = \begin{pmatrix} 1 & 2 & 3 & 3 \\ 2 & 0 & 6 & 2 \\ 3 & 4 & 9 & 7 \end{pmatrix}.$$

Then the column vectors are:

$$\begin{pmatrix} 1 \\ 2 \\ 3 \end{pmatrix}, \begin{pmatrix} 2 \\ 0 \\ 4 \end{pmatrix}, \begin{pmatrix} 3 \\ 6 \\ 9 \end{pmatrix}, \begin{pmatrix} 3 \\ 2 \\ 7 \end{pmatrix},$$

which form a vector space that has a basis of $\left\{ \begin{pmatrix} 1 \\ 2 \\ 3 \end{pmatrix}, \begin{pmatrix} 2 \\ 0 \\ 4 \end{pmatrix} \right\}$. As there are 2 vectors in this basis, the dimension of the column space is 2; hence, the rank of A is 2.

Nullspace

The nullspace or kernel of a matrix A (denoted ker (A)) is the set of all vectors x for which:

$$Ax = 0.$$

If x and y are in the nullspace, then $c_1 x + c_2 y$ is also in the nullspace as:

$$A(c_1 x + c_2 y) = c_1 (Ax) + c_2 (Ay) = 0 + 0 = 0,$$

so the nullspace is in fact a vector space. When *AA* is viewed as a linear transformation, the null-space is the subspace of $\mathbb{R}^n$ that is sent to 0 under the map A, hence the term "fundamental sub-space."

For example, consider the matrix:

$$A = \begin{pmatrix} 1 & 0 & 3 & 3 \\ 2 & 0 & 6 & 2 \\ 3 & 4 & 9 & 7 \end{pmatrix}.$$

The nullspace consists of all vectors x such that Ax = 0. This defines a system of linear equations that can be solved to give the family of solutions:

$$\begin{pmatrix} -x-3y \\ -x \\ y \\ x \end{pmatrix}, x, y \in \mathbb{R},$$

which defines a vector space with basis,

$$\left\{ \begin{pmatrix} -1 \\ -1 \\ 0 \\ 1 \end{pmatrix}, \begin{pmatrix} -3 \\ 0 \\ 1 \\ 0 \end{pmatrix} \right\}.$$

As there are two vectors in this basis, the dimension of the nullspace is 2.

Row Space and Left Nullspace

The row space and left nullspace are defined as the column space and nullspace of A^T, the transpose of A, respectively. In this way, they are subspaces of $\mathbb{R}^n$.

Fundamental Theorem of Linear Algebra

The fundamental theorem of linear algebra relates all four of the fundamental subspaces in a number of different ways. There are main parts to the theorem:

Part 1: The first part of the fundamental theorem of linear algebra relates the dimensions of the four fundamental subspaces:

The column and row spaces of an $m \times n$ matrix A *both* have dimension r, the rank of the matrix. The nullspace has dimension $n - r$, and the left nullspace has dimension $m - r$.

The dimension of the column space of:

$$A = \begin{pmatrix} 1 & 2 & 3 & 3 \\ 2 & 0 & 6 & 2 \\ 3 & 4 & 9 & 7 \end{pmatrix}$$

is 2, and the dimension of the nullspace of A is n - r = 4 - 2 = 2.

This first part of the fundamental theorem of linear algebra is sometimes referred to by name as the rank-nullity theorem.

Part 2: The second part of the fundamental theorem of linear algebra relates the fundamental subspaces more directly:

> Theorem: The nullspace and row space are orthogonal. The left nullspace and the column space are also orthogonal.

In other words, if v is in the nullspace of A and w is in the row space of A, the dot product $v \cdot w$ is 0. This is true because any vector in the nullspace is orthogonal to each row vector by definition, so it is also orthogonal to any linear combination of them.

Part 3: The final part of the fundamental theorem of linear algebra constructs an orthonormal basis, and demonstrates a singular value decomposition: any matrix M can be written in the form $U \sum V^T$.

where;

- U is an $m \times m$ unitary matrix;
- Σ is an m $m \times n$ matrix with nonnegative values on the diagonal;
- V is an $n \times n$ unitary matrix.

This part of the fundamental theorem allows one to immediately find a basis of the subspace in question.

This can be summarized in the following table:

Subspace	Subspace of	Symbol	Dimension	Basis
Column space	$\mathbb{R}^m$	im (A)	r = rank	First r columns of U
Nullspace (kernel)	$\mathbb{R}^n$	ker (A)	n − r	Last n − r columns of V
Row space	$\mathbb{R}^n$	im (AT)	r	First r columns of V
Left nullspace (kernel)	$\mathbb{R}^m$	ker (AT)	m − r	Last m − r columns of U

Or it can also be summarized in the following image, where the arrows represent the results of multiplication:

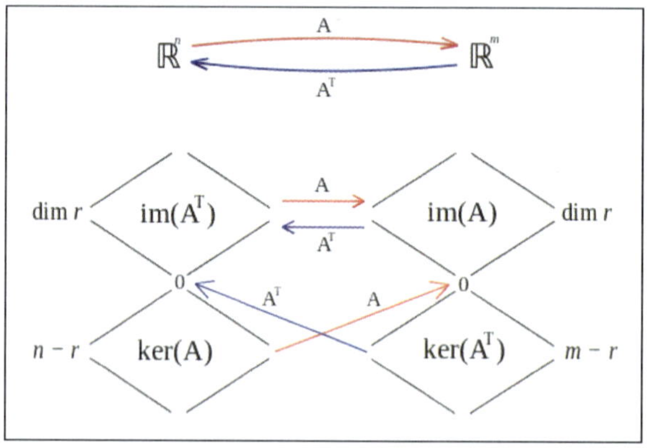

The relationship between the four fundamental subspaces.

References

- Linearalgebra: wolfram.com, Retrieved 13 February, 2019
- Introduction-to-matrices, boundless-algebra: lumenlearning.com, Retrieved 17 May, 2019
- Using-matrices-to-solve-systems-of-equations, boundless-algebra: lumenlearning.com, Retrieved 13 July, 2019
- Types-of-matrices, matrices, maths, guides: toppr.com, Retrieved 3 June, 2019
- Algebraicset: wolfram.com, Retrieved 19 May, 2019
- Determinant-of-a-matrix, determinants, maths, guides: toppr.com, Retrieved 28 April, 2019
- Reading-solving-systems-with-inverses, sanjacinto-finitemath1: lumenlearning.com, Retrieved 21 July, 2019
- 5-inverse-matrix.php top, matrices-determinants: intmath.com, Retrieved 15 January, 2019
- Adjoint-and-inverse-of-a-matrix, determinants, maths, guides: toppr.com, Retrieved 3 May, 2019

4

Elementary Algebra

The set of the basic concepts of algebra which are generally used for introducing variables is termed as elementary algebra. Some of the major concepts which are dealt with in elementary algebra are real numbers, exponents, polynomials and linear equations. This chapter discusses in detail these concepts related to elementary algebra.

Elementary algebra is the study of the arithmetic operators (+, -, *, /) and their rules along with the combinations of algebraic variables. If we combine some constants with some unknown variable, together with some arithmetic operations, then the expression is termed as an algebraic equation. The following are some of equations, representing algebraic equations or algebraic expressions.

Consider the following algebraic expression:

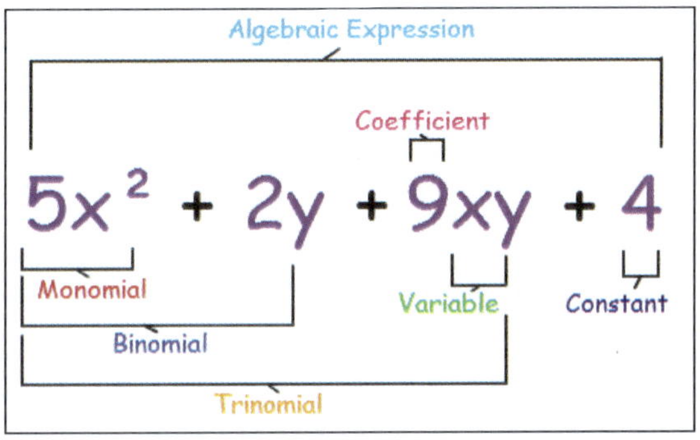

In the algebraic expression or polynomials, $5x^2$, 2y, 9xy, and 4 are known as terms.

Consider the term 9xy, here xy is the variable and the coefficient is 9. Similarly, in the term 2y, 2 is the coefficient and y is the variable.

By looking at the figure of algebraic expression we can clearly differentiate between these.

Elementary algebra includes many topics, out of which the most commonly used are mentioned below:

1. Factorization of $ax^2 + bx + c = 0$ (Quadratic equations) over the set of integers.
2. Operations with the algebraic fractions involving addition, subtraction, multiplication, and division.
3. Division of monomials and other polynomials including the simplification of algebraic fractions.
4. Solution to the linear equations and inequalities.
5. Solving the simultaneous system of linear equations.

6. Quadratic equations solution by the method of factoring.
7. The conversion of written phrases or sentences into algebraic expressions or equations.
8. Solving any verbal problem in an algebraic way including its geometrical reasoning through graphing.

Elementary Algebra Formulas

There are in numerous formulas in elementary algebra because we can derive many formulas from a given single formula. However, the following is the list of some of the important formulas used in elementary algebra:

a. Basic Laws of Elementary Algebra: Assuming that a and b are real numbers we have the following laws:

1. Closure Law: a + b and ab are real numbers.
2. Commutative Law: a + b = b + a, (for addition), ab = ba (for multiplication).
3. Associative Law: (a + b) + c = a + (b + c) (for addition), (ab) c = a (bc) (for multiplication).
4. Distributive Law: (a + b) c = ac + bc.
5. Identity Law: a + 0 = 0 + a = a.
6. Inverse Law: a + (-a) = 0, $a\left(\frac{1}{a}\right) = 1$, which means every number has its additive inverse.
7. Cancellation Law: If a + x = a + y, then, we can cancel a from both the sides, to get x = y.
8. Zero-factor Law: a * 0 = 0 * a = 0, multiplication with zero always gives the resulting answer as zero.
9. Negation Law: - (-a) = a, (-a) b = a(-b) = -(ab), (-a)(-b) = ab, negative of a negative number is a positive number and so on.

b. Factorization Formulas: The following formulas are used to factorize an algebraic expression according to their powers, as mentioned below:

1. $(a - b)(a^2 + ab + b^2) = a^3 - b^3$
2. $(a - b)^3 = a^3 - 3a^2b + 3ab^2 - b^3$
3. $(a + b)(a^2 - ab + b^2) = a^3 + b^3$
4. $(a + b)^3 = a^3 + 3a^2b + 3ab^2 + b^3$
5. $(a + b)(a - b) = a^2 - b^2$
6. $(a - b)^2 = a^2 - 2ab + b^2$
7. $(a + b)^2 = a^2 + 2ab + b^2$

c. Quadratic Equation formulas: A quadratic equation is an algebraic equation or expression written in the second degree, as shown below:

$$ax^2 + bx + c = 0,$$

where, a, b, c are the given numerical values of the coefficients, and x is an unknown variable. The roots of any quadratic equation can be found by any of the following methods:

1. The quadratic formula:

$$x = \frac{-b \pm \sqrt{b^2 - 4ac}}{2a}$$

2. Factoring Method: Here we split the middle term bx in such a way that the sum of the roots should be $\left(-\frac{b}{a}\right)$ and the product of the roots should be $\left(\frac{c}{a}\right)$.
3. Completing the Square Method: In this method we make a perfect square of c, by adding $(\frac{1}{2}$ of coefficient of $x)^2$ to both sides of the equation, to get a perfect or complete square.

d. Graphing of the quadratic equation: The discriminant of a quadratic function = $b^2 - 4ac$, where,

1. If the discriminant is positive, that is greater than zero, then it will have two real roots and the graph will have two x-intercepts.
2. If the discriminant is equal to zero, then the two roots will be real and equal to one another, and the graph will have only one x-intercept.
3. If the discriminant is less than zero, then the two roots will be imaginary and the graph will have no x-intercepts, and
   a. The graph will be drawn completely above the x-axis if $a > 0$.
   b. The graph will be drawn completely below the x-axis if $a < 0$.

Real Number

In Mathematics, real number is a quantity that can be expressed as an infinite decimal expansion. Real numbers are used in measurements of continuously varying quantities such as size and time, in contrast to the natural numbers 1, 2, 3, ... , arising from counting. The word *real* distinguishes them from the complex numbers involving the symbol *i*, or Square root of $\sqrt{-1}$, used to simplify the mathematical interpretation of effects such as those occurring in electrical phenomena. The real numbers include the positive and negative integers and fractions (or rational numbers) and also the irrational numbers. The irrational numbers have decimal expansions that do not repeat themselves, in contrast to the rational numbers, the expansions of which always contain a digit or group of digits that repeats itself, as 1/6 = 0.16666... or 2/7 = 0.285714285714.... The decimal formed as 0.42442444244442... has no regularly repeating group and is thus irrational.

The most familiar irrational numbers are algebraic numbers, which are the roots of algebraic equations with integer coefficients. For example, the solution to the equation $x^2 - 2 = 0$ is an algebraic irrational number, indicated by Square root of $\sqrt{2}$. Some numbers, such as π and e, are not the solutions of any such algebraic equation and are thus called transcendental irrational numbers. These numbers can often be represented as an infinite sum of fractions determined in some regular way, indeed the decimal expansion is one such sum.

The real numbers can be characterized by the important mathematical property of completeness, meaning that every nonempty set that has an upper bound has a smallest such bound, a property not possessed by the rational numbers. For example, the set of all rational numbers the squares of which are less than 2 has no smallest upper bound, because Square root of $\sqrt{2}$ is not a rational number. The irrational and rational numbers are both infinitely numerous, but the infinity of irrationals is "greater" than the infinity of rationals, in the sense that the rationals can be paired off with a subset of the irrationals, while the reverse pairing is not possible.

Set of Real Numbers

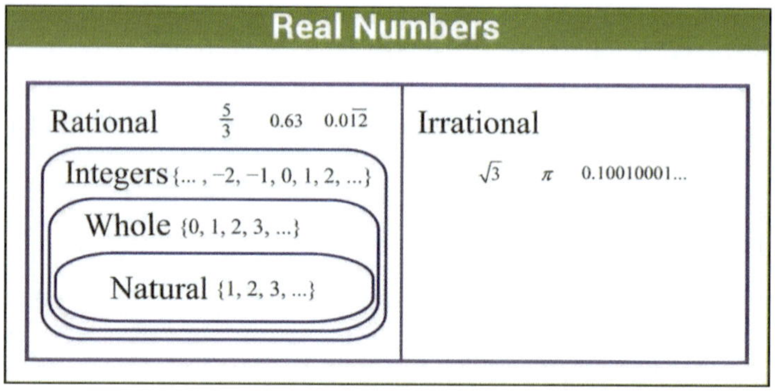

The set of real numbers are consisting of different categories, such as, natural and whole numbers, rational and irrational numbers and integers. In the table given here, all these numbers are defined with examples.

Category	Definition	Example
Natural Numbers	Contain all counting numbers which start from 1. N = {1, 2, 3, 4,......}	All numbers such as 1, 2, 3, 4, 5, 6, ...
Whole Numbers	Collection of zero and natural number. W = {0, ,2, 3,.....}	All numbers including 0 such as 0, 1, 2, 3, 4, 5, 6, ...
Integers	The collective result of whole numbers and negative of all natural numbers. All are real numbers.	Includes: -infinity,.........-4, -3, -2, -1, 0, 1, 2, 3, 4, ...+ infinity
Rational Numbers	Numbers that can be written in the form of p/q, where q ≠ 0.	Examples of rational numbers are ½, 5/4 and 12/6 etc.
Irrational Numbers	All the numbers which are not rational and cannot be written in the form of p/q.	Irrational numbers are non-terminating and non-repeating in nature like $\sqrt{2}$

Real Numbers Chart

The charts for the set of real numerals including all the types are given below:

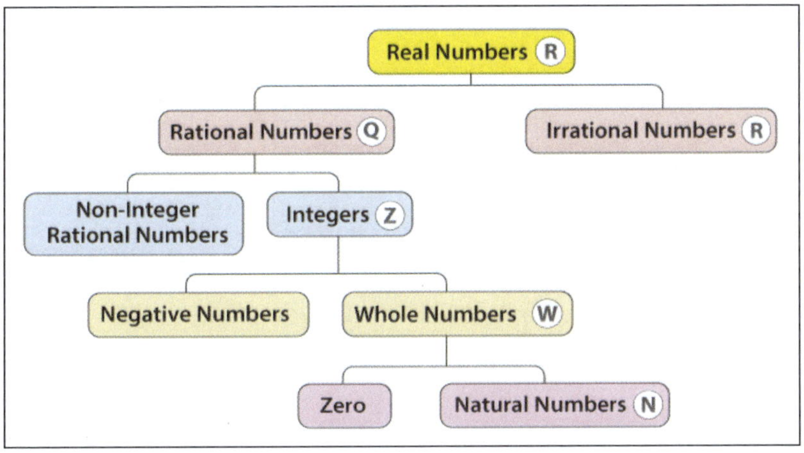

Properties of Real Numbers

There are four main properties which include commutative property, associative property, distributive property, and identity property. Consider "m, n and r" are the real numbers.

Commutative Property

If we m and n are the numbers, then the general form will be m + n = n + m for addition and m. n = n. m for multiplication.

- Addition: m + n = n + m. For example, 5 + 3 = 3 + 5, 2 + 4 = 4 + 2
- Multiplication: m × n = n × m. For example, 5 × 3 = 3 × 5, 2 × 4 = 4 × 2

Associative Property

If we m, n and r are the numbers. The general form will be m + (n + r) = (m + n) + r for addition (mn) r = m (nr) for multiplication.

- Addition: The general form will be m + (n + r) = (m + n) + r. An example of additive associative property is 10 + (3 + 2) = (10 + 3) + 2.
- Multiplication: (mn) r = m (nr). An example of a multiplicative associative property is (2 × 3) 4 = 2 (3 × 4).

Distributive Property

For three numbers m, n, and r, which are real in nature, the distributive property is in the form of:

m (n + r) = mn + mr and (m + n) r = mr + nr.

- Example of distributive property is: 5(2 + 3) = 5 × 2 + 5 × 3. Here, both sides will yield 25.

Identity Property

There are additive and multiplicative identities.

- For addition: m + 0 = m (zero is additive identity).
- For multiplication: a × 1 = 1 × a = a (1 is multiplicative identity).

Exponents

The base a raised to the power of n is equal to the multiplication of a, n times:

$$a^n = a \times a \times \dots \times a$$

n times a is the base and n is the exponent.

Examples:

$$3^1 = 3$$

$$3^2 = 3 \times 3 = 9$$

$$3^3 = 3 \times 3 \times 3 = 27$$

$$3^4 = 3 \times 3 \times 3 \times 3 = 81$$

$$3^5 = 3 \times 3 \times 3 \times 3 \times 3 = 243$$

Exponents Rules and Properties

Rule name	Rule	Example
Product rules	$a^n \cdot a^m = a^{n+m}$	$2^3 \cdot 2^4 = 2^{3+4} = 128$
	$a^n \cdot b^n = (a \cdot b)^n$	$3^2 \cdot 4^2 = (3 \cdot 4)^2 = 144$
Quotient rules	$a^n / a^m = a^{n-m}$	$2^5 / 2^3 = 2^{5-3} = 4$
	$a^n / b^n = (a / b)^n$	$4^3 / 2^3 = (4/2)^3 = 8$
Power rules	$(b^n)^m = b^{n \cdot m}$	$(2^3)^2 = 2^{3 \cdot 2} = 64$
	${}_b n^m {}_{= b}(n^m)$	${}_2 3^2 {}_{= 2}(3^2)_{= 512}$
	$\sqrt[m]{(b^n)} = b^{n/m}$	$\sqrt[2]{(2^6)} = 2^{6/2} = 8$
	$b^{1/n} = \sqrt[n]{b}$	$8^{1/3} = 3\ \sqrt{8} = 2$
Negative exponents	$b^{-n} = 1 / b^n$	$2^{-3} = 1/2^3 = 0.125$
Zero rules	$b^0 = 1$	$5^0 = 1$
	$0^n = 0$, for $n>0$	$0^5 = 0$
One rules	$b^1 = b$	$5^1 = 5$
	$1^n = 1$	$1^5 = 1$
Minus one rule	$(-1)^n = \begin{cases} 1 & ,n\ even \\ -1 & n\ odd \end{cases}$	$(-1)^5 = -1$

Derivative rule	$(x^n)' = n \cdot x^{n-1}$	$(x^3)' = 3 \cdot x^{3-1}$
Integral rule	$\int x^n dx = x^{n+1}/(n+1)+C$	$\int x^2 dx = x^{2+1}/(2+1)+C$

Addition and Subtraction of Exponents

To add or subtract with powers, both the variables and the exponents of the variables must be the same. You perform the required operations on the coefficients, leaving the variable and exponent as they are. When adding or subtracting with powers, the terms that combine always have exactly the same variables with exactly the same powers. These rules are true for multiplying and dividing exponents as well.

Examples:

1. $x + x + x = 3x$

   Because the variables are the same (x) and the powers are the same (there are no exponents, so the exponents must be 1), you can add the variables.

2. $x^2 - 2x^2 + 3x^2 + 3x^2 = 5x^2$

   Because the variables are the same (x) and the powers are the same (2), you can perform the required operations on the variables.

3. $4x^2 - 3x^3 + 2x^2 + x - 1$

   Although the variables are the same (x), the powers are not the same (1, 2, 3, and 4). You can't simplify these terms because only the variables are the same, and both the variables and the powers need to be the same.

4. $2x + 3x^2 + 5x^3 - 2x^2 - 3x - 1$

   $$5x^3 + \left(3x^2 - 2x^2\right) + \left(2x - 3x\right) - 1$$
   $$5x^3 + \left(3 - 2\right)x^2 + \left(2 - 3\right)x - 1$$
   $$5x^3 + x^2 - x - 1$$

Sometimes not all of the variables and powers will be the same — you may encounter a problem that has several groups of variables and powers that are not the same. In this case, you only add or subtract terms whose variables and powers are the same.

Multiplying Exponents

Multiplying Exponents with same Base

For exponents with the same base, we should add the exponents:

$$a^n \cdot a^m = a^{n+m}$$

Example: $2^3 \cdot 2^4 = 2^{3+4} = 2^7 = 2 \cdot 2 \cdot 2 \cdot 2 \cdot 2 \cdot 2 \cdot 2 = 128$

Multiplying Exponents with Different Bases

When the bases are different and the exponents of a and b are the same, we can multiply a and b first:

$a^n \cdot b^n = (a \cdot b)^n$

Example: $3^2 \cdot 4^2 = (3 \cdot 4)^2 = 12^2 = 12 \cdot 12 = 144$

When the bases and the exponents are different we have to calculate each exponent and then multiply:

$a^n \cdot b^m$

Example: $3^2 \cdot 4^3 = 9 \cdot 64 = 576$

Multiplying Negative Exponents

For exponents with the same base, we can add the exponents:

$a^{-n} \cdot a^{-m} = a^{-(n+m)} = 1 / a^{n+m}$

Example: $2^{-3} \cdot 2^{-4} = 2^{-(3+4)} = 2^{-7} = 1 / 2^7 = 1 / (2 \cdot 2 \cdot 2 \cdot 2 \cdot 2 \cdot 2 \cdot 2) = 1 / 128 = 0.0078125$

When the bases are different and the exponents of a and b are the same, we can multiply a and b first:

$a^{-n} \cdot b^{-n} = (a \cdot b)^{-n}$

Example: $3^{-2} \cdot 4^{-2} = (3 \cdot 4)^{-2} = 12^{-2} = 1 / 12^2 = 1 / (12 \cdot 12) = 1 / 144 = 0.0069444$

When the bases and the exponents are different we have to calculate each exponent and then multiply:

$a^{-n} \cdot b^{-m}$

Example: $3^{-2} \cdot 4^{-3} = (1/9) \cdot (1/64) = 1 / 576 = 0.0017361$

Multiplying Fractions with Exponents

Multiplying fractions with exponents with same fraction base:

$(a / b)^n \cdot (a / b)^m = (a / b)^{n+m}$

Example: $(4/3)^3 \cdot (4/3)^2 = (4/3)^{3+2} = (4/3)^5 = 4^5 / 3^5 = 4.214$

Multiplying fractions with exponents with same exponent:

$(a / b)^n \cdot (c / d)^n = ((a / b) \cdot (c / d))^n$

Example: $(4/3)^3 \cdot (3/5)^3 = ((4/3) \cdot (3/5))^3 = (4/5)^3 = 0.8^3 = 0.8 \cdot 0.8 \cdot 0.8 = 0.512$

Multiplying fractions with exponents with different bases and exponents:

$(a / b)^n \cdot (c / d)^m$

Example: $(4/3)^3 \cdot (1/2)^2 = 2.37 \cdot 0.25 = 0.5925.$

Multiplying Fractional Exponents

Multiplying fractional exponents with same fractional exponent:

$$a^{n/m} \cdot b^{n/m} = (a \cdot b)^{n/m}$$

Example: $2^{3/2} \cdot 3^{3/2} = (2{\cdot}3)^{3/2} = 6^{3/2} = \sqrt{(6^3)} = \sqrt{216} = 14.7$

Multiplying fractional exponents with same base:

$$a^{(n/m)} \cdot a^{(k/j)} = a^{[(n/m)+(k/j)]}$$

Example: $2^{(3/2)} \cdot 2^{(4/3)} = 2^{[(3/2)+(4/3)]} = 7.127$

Multiplying fractional exponents with different exponents and fractions:

$a^{n/m} \cdot b^{k/j}$

Example: $2^{3/2} \cdot 2^{4/3} = \sqrt{(2^3)} \cdot \sqrt[3]{(2^4)} = 2.828 \cdot 2.52 = 7.127$

Multiplying Square Roots with Exponents

For exponents with the same base, we can add the exponents:

$$(\sqrt{a})^n \cdot (\sqrt{a})^m = a^{(n+m)/2}$$

Example: $(\sqrt{5})^2 \cdot (\sqrt{5})^4 = 5^{(2+4)/2} = 5^{6/2} = 5^3 = 125$

Multiplying Variables with Exponents

For exponents with the same base, we can add the exponents:

$$x^n \cdot x^m = x^{n+m}$$

Example: $x^2 \cdot x^3 = (x{\cdot}x) \cdot (x{\cdot}x{\cdot}x) = x^{2+3} = x^5$

Dividing Exponents

Dividing Exponents with same Base

For exponents with the same base, we should subtract the exponents:

$$a^n / a^m = a^{n-m}$$

Example: $2^6 / 2^3 = 2^{6-3} = 2^3 = 2{\cdot}2{\cdot}2 = 8$

Dividing Exponents with Different Bases

When the bases are different and the exponents of a and b are the same, we can divide a and b first:

$$a^n / b^n = (a / b)^n$$

Example: $6^3 / 2^3 = (6/2)^3 = 3^3 = 3 \cdot 3 \cdot 3 = 27$

When the bases and the exponents are different we have to calculate each exponent and then divide:

$$a^n / b^m$$

Example: $6^2 / 3^3 = 36 / 27 = 1.333$

Dividing Negative Exponents

For exponents with the same base, we can subtract the exponents:

$$a^{-n} / a^{-m} = a^{-n-(-m)} = a^{m-n}$$

Example: $2^{-3} / 2^{-5} = 2^{5-3} = 2^2 = 2 \cdot 2 = 4$

When the bases are different and the exponents of a and b are the same, we can multiply a and b first:

$$a^{-n} / b^{-n} = (a/b)^{-n} = 1 / (a/b)^n = (b/a)^n$$

Example: $3^{-2} / 4^{-2} = (4/3)^2 = 1.7778$

When the bases and the exponents are different we have to calculate each exponent and then divide:

$$a^{-n} / b^{-m} = b^m / a^n$$

Example: $3^{-2} / 4^{-3} = 4^3 / 3^2 = 64 / 9 = 7.111$

Dividing Fractions with Exponents

Dividing fractions with exponents with same fraction base:

$(a / b)^n / (a / b)^m = (a / b)^{n-m}$

Example: $(4/3)^3 / (4/3)^2 = (4/3)^{3-2} = (4/3)^1 = 4/3 = 1.333$

Dividing fractions with exponents with same exponent:

$$(a / b)^n / (c / d)^n = ((a / b)/(c / d))^n = ((a \cdot d / b \cdot c))^n$$

Example: $(4/3)^3 / (3/5)^3 = ((4/3)/(3/5))^3 = ((4 \cdot 5)/(3 \cdot 3))^3 = (20/9)^3 = 10.97$

Dividing fractions with exponents with different bases and exponents:

$$(a / b)^n / (c / d)^m$$

Example: $(4/3)^3 / (1/2)^2 = 2.37 / 0.25 = 9.481$

Dividing Fractional Exponents

Dividing fractional exponents with same fractional exponent:

$$a^{n/m} / b^{n/m} = (a / b)^{n/m}$$

Example: $3^{3/2} / 2^{3/2} = (3/2)^{3/2} = 1.5^{3/2} = \sqrt{(1.5^3)} = \sqrt{3.375} = 1.837$

Dividing fractional exponents with same base:

$$a^{n/m} / a^{k/j} = a^{(n/m)-(k/j)}$$

Example:

$$2^{3/2} / 2^{4/3} = 2^{(3/2)-(4/3)} = 2^{(1/6)} = \sqrt[6]{2} = 1.122$$

Dividing fractional exponents with different exponents and fractions:

$$a^{n/m} / b^{k/j}$$

Example: $2^{3/2} / 2^{4/3} = \sqrt{(2^3)} / \sqrt[3]{(2^4)} = 2.828 / 2.52 = 1.1222$

Dividing Variables with Exponents

For exponents with the same base, we can subtract the exponents:

$$x^n / x^m = x^{n-m}$$

Example:

$$x^5 / x^3 = (x \cdot x \cdot x \cdot x \cdot x) / (x \cdot x \cdot x) = x^{5-3} = x^2$$

Polynomials

A polynomial is an expression containing two or more algebraic terms. They are often the sum of several terms containing different powers (exponents) of variables.

Examples of Polynomials

In a polynomial, terms consist of variables or constants (numbers) that are multiplied, added, or subtracted from each other. Additionally, any exponents must be positive whole numbers. For example, each of the expressions below would qualify as a polynomial:

- $5x^2$
- $2x^4 - 5x + 11$
- $\frac{2}{3}z^5 + 2xyz - xy$

In each case above, notice that the exponents were positive integers (positive whole numbers). If an exponent is negative, then this implies you are dividing by that term, and polynomials cannot have any division involving the variables. A polynomial can have fractions involving just the numbers in front of the variables (the coefficients), but not involving the variables.

Examples of Expressions which are not Polynomials

Keeping the explanation above in mind, the following are not polynomials:

- $x^{-2} + x - 1$

  There is a negative exponent.

- xy^{-}

  The exponent is a fraction.

- $\log(x) + 3x^4$

  This involves another function: log(x).

Special Types of Polynomials

There are a few special types of polynomials which get their own names. Remember, we are thinking of a term as something like 3xy or 5. In other words, it is one of the pieces of the polynomial that is being added or subtracted from the other pieces/terms. When counting terms, you are assuming that none of the terms are like terms.

Monomials

Monomials are polynomials with a single term. Some examples are:

- $-3x^4$
- 10xy
- 2x

Binomials

Binomials have two terms. One of the most common types of binomials you will see in algebra are those like 3x + 5 or x + 4. These involve one variable term and one constant term and are often the factors found when working with quadratic equations.

Some other examples of binomials are:

- $\frac{1}{2}x^5 + 2x$
- $4x - 9$
- $12x^2y - 4xy^2$

Trinomials

Trinomials have three terms. One place you will see these lot is the classic "quadratic equation" like $2x^2 + 5x + 1 = 0$. The expression on the left-hand side of this equation is a trinomial.

Some other examples of trinomials are:

- $-1 + 4x + 3y$
- $2x^2 y + 3x + 7$
- $x^2 + 3x + 1$

Elements of a Polynomial

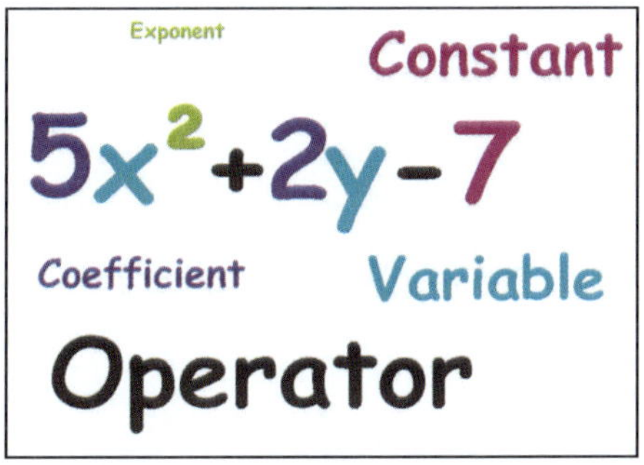

A polynomial can contain variables, constants, coefficients, exponents, and operators.

A polynomial is an algebraic expression made up of two or more terms. Polynomials can be made up of some or all of the following:

- Variables: These are letters like x, y, and b.
- Constants: These are numbers like 3, 5, 11. They are sometimes attached to variables, but can also be found on their own.
- Exponents: Exponents are usually attached to variables, but can also be found with a constant. Examples of exponents include the 2 in 5^2 or the 3 in x^3.
- Addition, subtraction, multiplication, and division: For example, you can have 2x (multiplication), 2x+5 (multiplication and addition), and x-7 (subtract.)

What isn't a Polynomial

There are a few rules as to what polynomials cannot contain:

- Polynomials cannot contain division by a variable: For example, $2y^2+7x/4$ is a polynomial, because 4 is not a variable. However, 2y2+7x/(1+x) is not a polynomial as it contains division by a variable.
- Polynomials cannot contain negative exponents: You cannot have $2y^{-2}+7x-4$. Negative exponents are a form of division by a variable (to make the negative exponent positive, you have to divide.) For example, x^{-3} is the same thing as $1/x^3$.
- Polynomials cannot contain fractional exponents: Terms containing fractional exponents (such as $3x+2y^{1/2}-1$) are not considered polynomials.
- Polynomials cannot contain radicals: For example, $2y^2 +\sqrt{3}x + 4$ is not a polynomial.

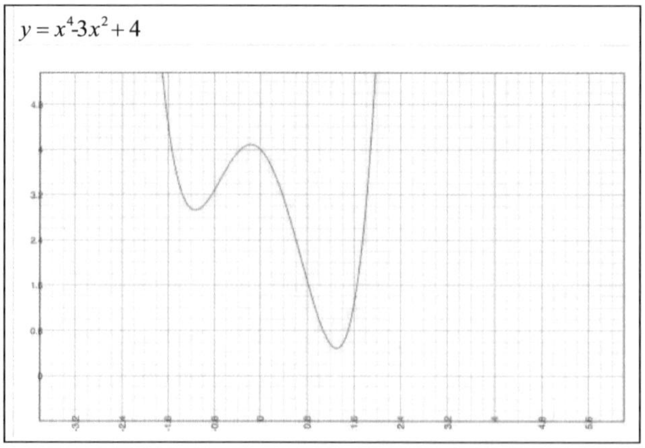

A graph of a polynomial of a single variable shows nice curvature.

Degree of a Polynomial

To find the degree of a polynomial, write down the terms of the polynomial in descending order by the exponent. The term whose exponents add up to the highest number is the leading term. The sum of the exponents is the degree of the equation.

Example: Figure out the degree of $7x^2y^2+5y^2x+4x^2$.

Start out by adding the exponents in each term.

The exponents in the first term, $7x^2y^2$ are 2 (from $7x^2$) and 2 (from y^2) which add up to four.

The second term ($5y^2x$) has two exponents. They are 2 (from $5y^2$) and 1 (from x, this is because x is the same as x^1.) The exponents in this term add up to three.

The last term ($4x^2$) only has one exponent, 2, so its degree is just two.

Since the first term has the highest degree (the 4th degree), it is the leading term. The degree of this polynomial is four.

Zeroes of Polynomial

A polynomial having value zero (0) is known as zero polynomial. Actually, the term 0 is itself zero polynomial. It is a constant polynomial whose all the coefficients are equal to 0. For a polynomial, there may be few (one or more) values of the variable for which the polynomial may result in zero. These values are known as zeros of a polynomial. We can say that the zeroes of a polynomial are defined as the points where the polynomial equals to zero on the whole.

Ifthecoefficientsoffollowingtheformofthepolynomial: $a_nx^n + a_{n-1}x^{n-1} + a_{n-2}x^{n-2} + ... + a_2x^2 + a_1x + a_0$ are zero, then it will become zero polynomial. i.e. $a_n = a_{n-1} = a_{n-2} = ... = a_0 = 0.$. Thus, the polynomial will become 0 and may be written as P(x) = 0.

Zero Polynomial Function

The zero polynomial function is defined as the polynomial function with the value of zero. i.e. the function whose value is 0, is termed as a zero polynomial function. Zero polynomial does not have any nonzero term. It is represented as: P(x) = 0. Thus, we can say that a polynomial function which is equal to zero is called zero polynomial function. It also is known as zero maps. The graph of the zero polynomial is X axis.

Examples:

- Let's look at the polynomial, $p(x) = 5x^3 - 2x^2 + 3x - 2$. Now, let's find the value of the polynomial(x) at x = 1, $p(1) = 5(1)^3 - 2(1)^2 + 3(1) - 2 = 5 - 2 + 3 - 2 = 4$. Therefore, we can say that the value of the polynomial p(x) at x = 1 is 4.

  Next, let's find the value of the polynomial(x) at x = 0, $p(0) = 5(0)^3 - 2(0)^2 + 3(0) - 2 = 0 - 0 + 0 - 2 = -2$. Therefore, we can say that the value of the polynomial p(x) at x = 0 is – 2.

- 2. Let's look at another polynomial now, p(x) = x – 1. Let's find the value of the polynomial at x = 1, p(1) = 1 – 1 = 0. So, the value of the polynomial p(x) at x = 1 is 0. Therefore, 1 is the zero of the polynomial p(x).

Similarly, for the polynomial q(x) = x – 2, the zero of the polynomial is 2. To summarize, zeroes of polynomial p(x) are numbers c and d such that p(c) = 0 and p(d) = 0.

Calculating Zeroes of Polynomial

When we calculated zeroes of polynomial p(x) = x – 1, we equated it to 0, x – 1 = 0 or, x = 1. Hence, we say that p(x) = 0 is the Polynomial Equation. 1 is the Root of the Polynomial equation p(x) = 0. OR 1 is the Zero of the Polynomial equation p(x) = x – 1 = 0.

Now, let's look at a constant polynomial '5'. You can write this as $5x^0$. What is the Root of this constant polynomial? The answer is a Non-zero constant polynomial has no zero. Also, every real number is a zero of the Zero Polynomial.

Let's look at the following linear polynomial to understand the calculation of the roots or 'zeroes of polynomial': p(x) = ax + b ... where a ≠ 0. To find a zero, we must equate the polynomial to 0. [p(x) = 0]. Hence, ax + b = 0 ... where a ≠ 0. So, ax = – b or, x = – b/a. Therefore, '- b/a' is the only zero of p(x) = ax + b. We can also say that a linear polynomial has only one zero.

Observations

- A zero of a polynomial need not be 0.
- 0 may be a zero of a polynomial.
- Every linear polynomial has one and only one zero.
- A polynomial can have more than one zero.

Solved Examples:

- Find p(0), p(1) and p(2) for each of the following polynomials:
  - $p(y) = y^2 - y + 1$
  - $p(t) = 2 + t + 2t^2 - t^3$

- $p(x) = x^3$
- $p(x) = (x - 1)(x + 1)$

Solution:

$p(y) = y^2 - y + 1$

- $p(0) = 0^2 - 0 + 1 = 1$
- $p(1) = 1^2 - 1 + 1 = 1$
- $p(2) = 2^2 - 2 + 1 = 3$

$p(t) = 2 + t + 2t^2 - t^3$

- $p(0) = 2 + 0 + 2(0)^2 - (0)^3 = 2 + 0 + 0 - 0 = 2$
- $p(1) = 2 + 1 + 2(1)^2 - (1)^3 = 2 + 1 + 2 - 1 = 4$
- $p(2) = 2 + 2 + 2(2)^2 - (2)^3 = 2 + 2 + 8 - 8 = 4$

$p(x) = x^3$

- $p(0) = (0)^3 = 0$
- $p(1) = (1)^3 = 1$
- $p(2) = (2)^3 = 8$

$p(x) = (x - 1)(x + 1)$

- $p(0) = (0 - 1)(0 + 1) = (-1)(1) = -1$
- $p(1) = (1 - 1)(1 + 1) = (0)(2) = 0$
- $p(2) = (2 - 1)(2 + 1) = (1)(3) = 3$

Verify the following:

- Verify whether the following are zeroes of polynomials indicated against them:
  - $p(x) = 3x + 1, x = -1/3$
  - $p(x) = 5x - \pi, x = 4/5$
  - $p(x) = (x + 1)(x - 2), x = -1, 2$
  - $p(x) = 3x^2 - 1, x = -1/\sqrt{3}, 2/\sqrt{3}$

Solution:

- $p(x) = 3x + 1, x = -1/3$
- $p(-1/3) = 3(-1/3) + 1 = -1 + 1 = 0.$

Hence, $x = -1/3$ is a zero of polynomial $3x + 1$.

- $p(x) = 5x - \pi, x = 4/5$
- $p(4/5) = 5(4/5) - \pi = 4 - \pi \neq 0.$

Hence, $x = 4/5$ is not a zero of polynomial $5x - \pi$.

- $p(x) = (x + 1)(x - 2), x = -1, 2$
- $p(-1) = (-1 + 1)(-1 - 2) = (0)(-3) = 0.$

Hence, $x = -1$ is a zero of polynomial $(x + 1)(x - 2)$.

- $p(2) = (2 + 1)(2 - 2) = (3)(0) = 0.$

Hence, $x = 2$ is a zero of polynomial $(x + 1)(x - 2)$.

- $p(x) = 3x^2 - 1, x = -1/\sqrt{3}, 2/\sqrt{3}$
- $p(-1/\sqrt{3}) = 3(-1/\sqrt{3})^2 - 1 = 3(1/3) - 1 = 1 - 1 = 0.$

Hence, $x = -1/\sqrt{3}$ is a zero of polynomial $3x^2 - 1$.

- $p(2/\sqrt{3}) = 3(2/\sqrt{3}) - 1 = 3(4/3) - 1 = 4 - 1 = 3 \neq 0.$

Hence, $x = 2/\sqrt{3}$ is not a zero of polynomial $3x^2 - 1$.

Geometrical Representation of Zeroes of a Polynomial

Linear Polynomial

Let's look at a linear polynomial ax + b, where a ≠ 0. The graph of y = ax + b is a straight line. Let's look at the graph of y = 2x + 3.

x	-2	2
y = 2x + 3	-1	7

The straight line y = 2x + 3 will pass through the points (– 2, – 1) and (2, 7). Here is how the graph looks like:

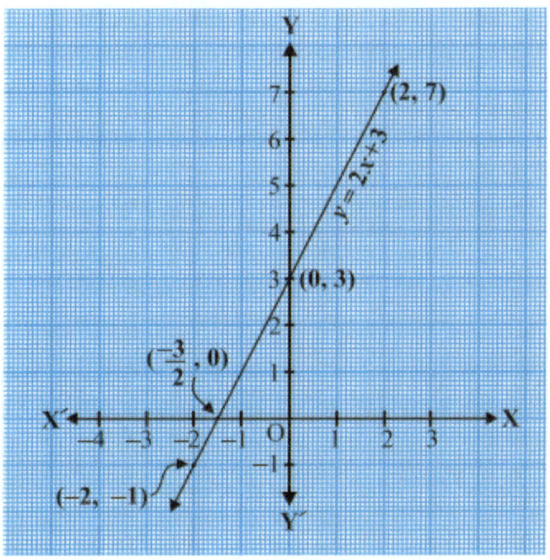

From the figure, you can see that the graph of $y = 2x + 3$ intersects the x-axis at the point $(-3/2, 0)$. Now, the zero of $(2x + 3)$ is $(-3/2)$. Therefore, the zero of the linear polynomial $(2x + 3)$ is the x-coordinate of the point where the graph of $y = 2x + 3$ intersects the x-axis. Hence, we can say,

"For a linear polynomial $ax + b$, $a \neq 0$, the graph of $y = ax + b$ is a straight line which intersects the x-axis at exactly one point, namely, $(-b/a, 0)$. Also, this linear polynomial has only one zero which is the x-coordinate of the point where the graph of $y = ax + b$ intersects the x-axis."

Quadratic Polynomial

Let's look at a quadratic polynomial, $x^2 - 3x - 4$. To look at the graph of $y = x^2 - 3x - 4$, let's list some values:

x	-2	-1	0	1	2	3	4	5
$y = x^2 - 3x - 4$	6	0	-4	-6	-6	-4	0	6

The graph of $y = x^2 - 3x - 4$ will pass through $(-2, 6)$, $(-1, 0)$, $(0, -4)$, $(1, -6)$, $(2, -6)$, $(3, -4)$, $(4, 0)$ and $(5, 6)$. Here is how the graph looks:

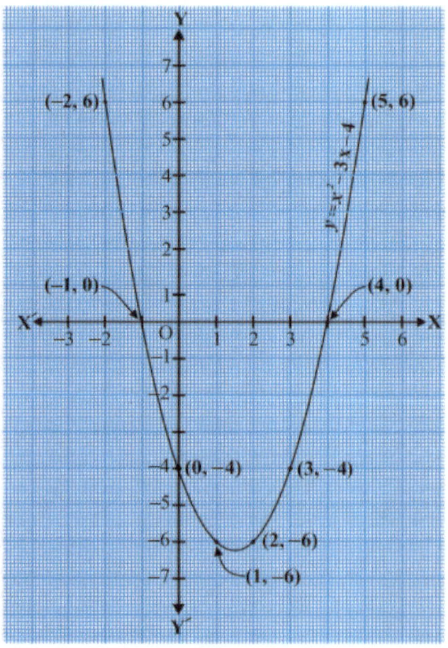

For that matter, for any quadratic polynomial $y = ax^2 + bx + c$, $a \neq 0$, the graph of $y = ax^2 + bx + c$ has either one of these two shapes:

- If $a > 0$, then it is open upwards like the one shown in the graph.
- If $a < 0$, then it is open downwards.

These curves are parabolas. A quick look at the table above shows that (-1) and (4) are zeroes of the quadratic polynomial. From the figure, you can see that (-1) and (4) are the x-coordinates of the points where the graph of $y = x^2 - 3x - 4$ intersects the x-axis. Therefore, we can say,

"The zeroes of a quadratic polynomial $ax^2 + bx + c$, $a \neq 0$, are precisely the x-coordinates of the points where the parabola representing $y = ax^2 + bx + c$ intersects the x-axis."

As far as the shape of the graph is concerned, the following three cases are possible:

Case (i): The graph cuts x-axis at two distinct points A and A′, where the x-coordinates of A and A′ are the two zeroes of the quadratic polynomial $ax^2 + bx + c$, as shown below:

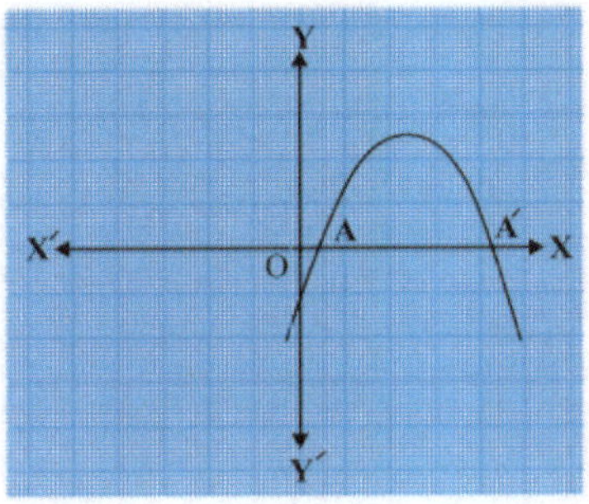

(i)

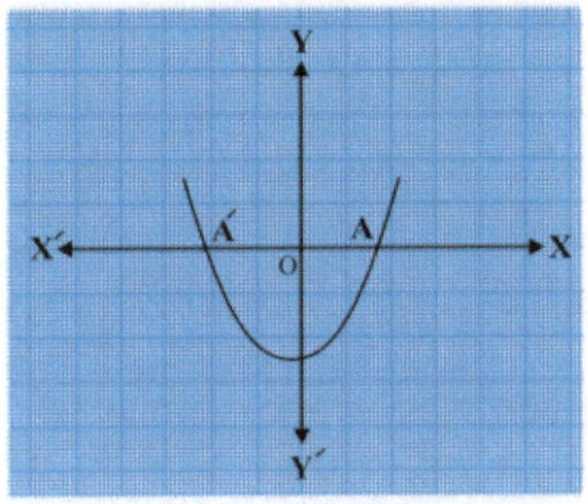

(ii)

Case (ii): The graph intersects the x-axis at only one point, or at two coincident points. Also, the x-coordinate of A is the only zero for the quadratic polynomial ax2 + bx + c, as shown below:

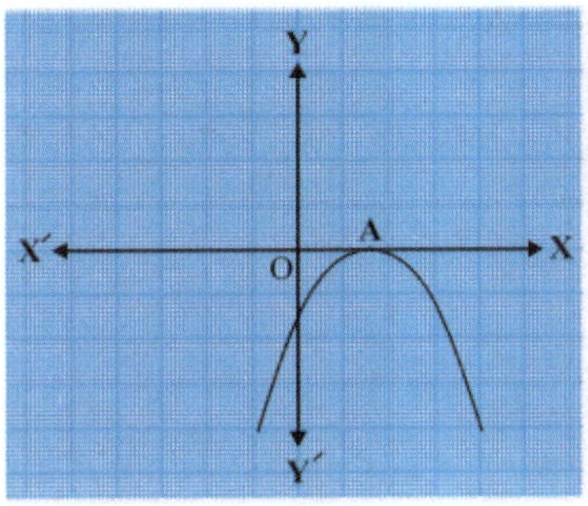

(i)

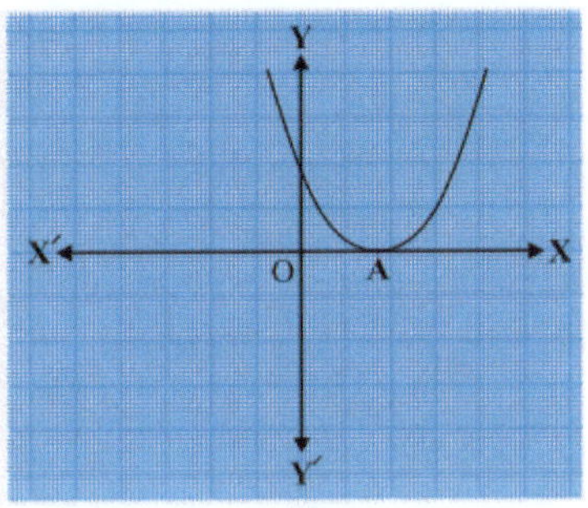

(ii)

Case (iii): The graph is either:

- Completely above the x-axis, or
- Completely below the x-axis.

So, it does not cut the x-axis at any point. Hence, the quadratic polynomial $ax^2 + bx + c$ has no zero, as shown below:

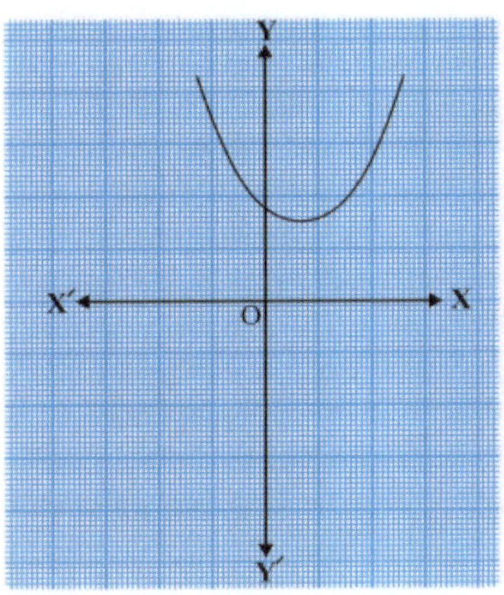

(i)

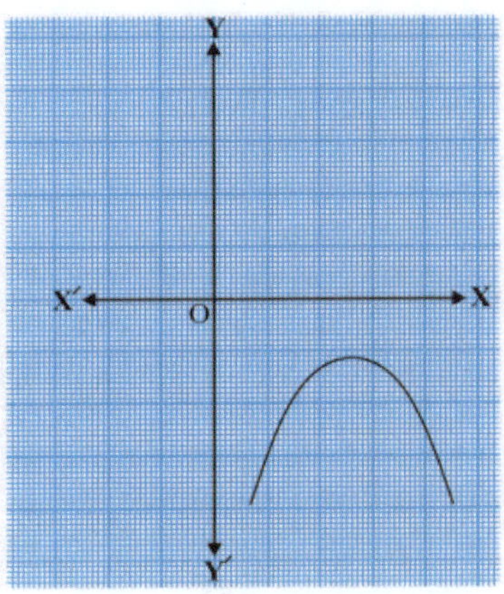

(ii)

To summarize, a quadratic polynomial can have either:

- Two distinct zeroes (as shown in Case i).
- Two equal zeroes (or one zero as shown in Case ii).
- No zero (as shown in Case iii).

It can also be summarized by saying that a polynomial of degree 2 has a maximum of 2 zeroes.

Cubic Polynomial

Let's look at a cubic polynomial, $x^3 - 4x$. Next, let's list a few values to plot the graph of $y = x^3 - 4x$.

x	− 2	− 1	0	1	2
$y = x^3 - 4x$	0	3	0	− 3	0

The graph of $y = x^3 - 4x$ will pass through (- 2, 0), (- 1, 3), (0, 0), (1, − 3), and (2, 0). Here is how the graph looks like:

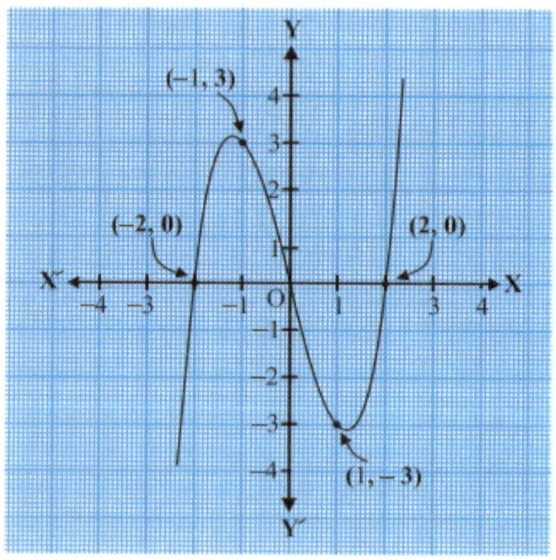

From the table, we can see that 2, 0 and − 2 are the zeroes of the cubic polynomial $x^3 - 4x$. You can also observe that the graph of $y = x^3 - 4x$ intersects the x-axis at 2, 0 and − 2. Let's take a quick look at some examples:

Let's plot the graph of the following two cubic polynomials:

- x^3
- $x^3 - x^2$

The graphs of $y = x^3$ *and* $y = x^3 - x^2$ *look as follows:*

From the first graph, you can observe that 0 is the only zero of the polynomial x^3, since the graph of $y = x^3$ intersects the x-axis only at 0. Similarly, the polynomial $x^3 - x^2 = x^2(x - 1)$ has two zeroes, 0 and 1. From the second diagram, you can see that the graph of $y = x^3 - x^2$ intersects the x-axis at 0 and 1.

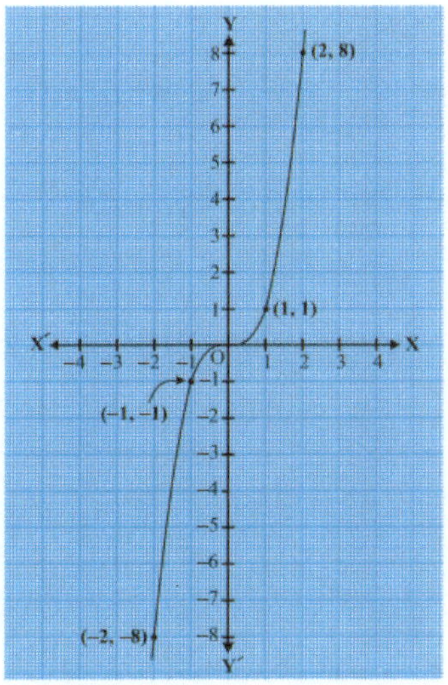

$y = x^3$

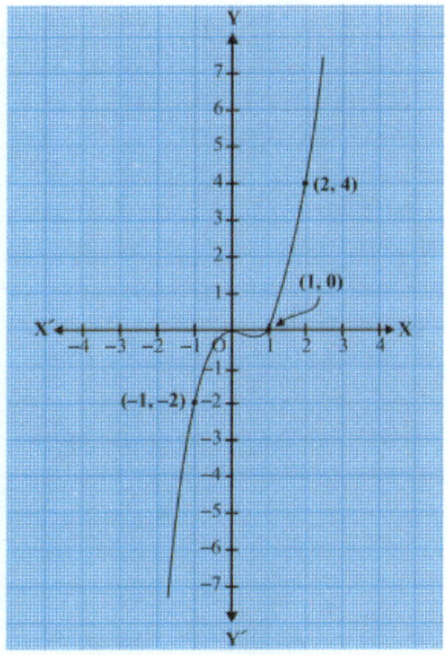

$y = x^3 - x^2$

Hence, we can conclude that there is a maximum of three zeroes for any cubic polynomial. Or, any polynomial with degree 3 can have maximum 3 zeroes. In general, given a polynomial p(x) of degree n, the graph of y = p(x) intersects the x-axis at a maximum of n points. Therefore, a polynomial p(x) of degree n has a maximum of n zeroes.

Example: The graphs of y = p(x) are given in the figure below, for some polynomials p(x). Find the number of zeroes of p(x), in each case.

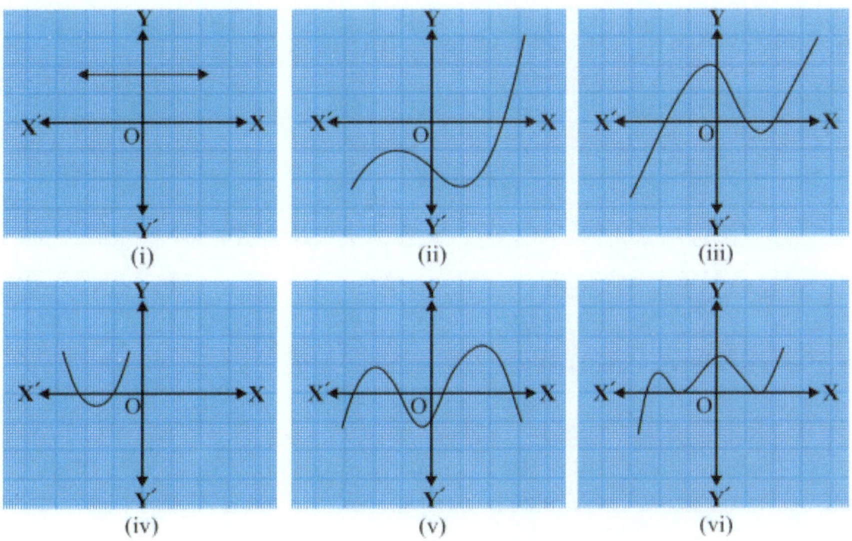

Solution: The numbers of zeroes in each of the graphs above are,

- 0, since the line is not intersecting the x-axis.
- 1, since the line intersects the x-axis once.
- 2, since the line intersects the x-axis twice.
- 2, since the line intersects the x-axis twice.

- 4, since the line intersects the x-axis four times.
- 3, since the line intersects the x-axis thrice.

Linear Equation

Linear Equation is a statement that a first-degree polynomial—that is, the sum of a set of terms, each of which is the product of a constant and the first power of a variable—is equal to a constant. Specifically, a linear equation in n variables is of the form $a_0 + a_1x_1 + ... + a_nx_n = c$, in which $x_1, ..., x_n$ are variables, the coefficients $a_0, ..., a_n$ are constants, and c is a constant. If there is more than one variable, the equation may be linear in some variables and not in the others. Thus, the equation $x + y = 3$ is linear in both x and y, whereas $x + y^2 = 0$ is linear in x but not in y. Any equation of two variables, linear in each, represents a straight line in Cartesian coordinates; if the constant term $c = 0$, the line passes through the origin.

A set of equations that has a common solution is called a system of simultaneous equations. For example, in the system:

$$2x - y = 1$$
$$x + 2y = 8$$

Both equations are satisfied by the solution x = 2, y = 3. The point (2, 3) is the intersection of the straight lines represented by the two equations.

A linear differential equation is of first degree with respect to the dependent variable (or variables) and its (or their) derivatives. As a simple example, note $dy/dx + Py = Q$, in which P and Q can be constants or may be functions of the independent variable, x, but do not involve the dependent variable, y. In the special case that P is a constant and $Q = 0$, this represents the very important equation for exponential growth or decay (such as radioactive decay) whose solution is $y = ke^{-Px}$, where e is the base of the natural logarithm.

Linear Equations in One Variable

The expressions with only one variable are called Linear Equations in one variable. These expressions have only one variable. Variables are the numbers mentioned in terms of any alphabets, in the examples given we see variables like a, b, c, x, y.

Solving Various Types of Linear Equations in One Variable

Solving Linear equations in one variable involves a few techniques of balancing the two expressions on both sides of the equal sign.

Technique 1: For understanding this technique let's consider the example given below: $4x - 5 = 15$, adding 5 to both sides:

$$4x - 5 + 5 = 15 + 5$$

$$4x = 20$$

Now, divide both sides with 4,

$$4x/4 = 20/4$$

$$x = 5$$

Technique 2: The following example will help us understand the technique in a better way: $3z + 8 = 10$, changing the position of 8 to R.H.S,

$$3z = 10 - 8$$

When we move a number from one side to another, the sign before the number inverts to the opposite sign. – sign becomes +; × sign becomes ÷ and vice-versa. Now,

$$3z = 2$$

$$\text{so, } z = 2/3$$

Let's check our answer by equating the two sides and replacing the solution with the variable.

$$3z + 8 = 10$$

$$3(2/3) + 8 = 10$$

$$2 + 8 = 10$$

$$10 = 10$$

L.H.S = R.H.S, hence the value of *z* is accurate.

Solving Linear Equations having Variables on Both Sides

For solving such equations, the techniques used above shall be followed, but with care and accuracy of operations. The following examples will make our understanding of the same clear:

$$2a - 5 = a + 4$$

$$2a = a + 4 + 5$$

or,

$$2a = a + 9$$

Now, we subtract a on both sides;

$$2a - a = a + 9 - a$$

$$a = 9$$

The point to be noted here in the example above is that we have subtracted a variable "a" from both sides. In an algebraic equation, a variable is also considered a number as it in a way represents a number. Let's take another example:

$$3y + 5/2 = 3y/2 - 14$$

Let's multiply both sides with 2 (as it is the denominator on both sides),

$$2\times (3y + 5/2) = 2\times (3y/2 - 14)$$

and $(2 \times 3y) + 2 \times (5/2) = (2 \times 3y/2) - (2 \times 14)$

$$6y + 5 = 3y - 28$$

We change the position of 3y from RHS to LHS,

$$6y - 3y + 5 = - 28$$

$$3y + 5 = - 28$$

$$3y = - 28 - 5$$

$$y = - 33/3$$

$$y = - 11$$

Reducing a Linear Equation to Simpler Forms

Linear equations may not always be simple. We may come up front with equations that are a combination of many operations. In such a case, we apply the basic rule of B.O.D.M.A.S. The following example will make things clear:

$$25a - 4(4a - 3) = 4(3a - 2) + 7/2$$

The first step here involves opening of all the brackets. Let's take the expression on LHS:

$$25a - 4(4a - 3)$$

$$= 25a - 16a + 12$$

$$=9a + 12$$

Now let's take RHS,

$$4(3a - 2) + 7/2$$

$$=12a - 8 + 7/2$$

$$=12a - 16/2 + 7/2$$

So, = 12a – 9/2

Now the equation is; which includes the simplified expression of LHS and RHS:

$9a + 12 = 12a - 9/2$

$12 = 12a - 9a - 9/2$

$12 = 3a - 9/2$

$12+9/2 = 3a$

$24/2 + 9/2 = 3a$

$33/2 = 3a$

$33/2 \times 1/3 = a$

$a = 11/2$

Now, let's check our solution by replacing the solution with the variable:

$25a - 4(4a - 3) = 4(3a - 2) + 7/2$

$LHS = 25 \times 11/2 - 4(4 \times 11/2 - 3)$

$= 275/2 - 4(22 - 3)$

$= 275/2 - 76 = 275/2 - 152/2 = 123/2$

$RHS = 4(3a - 2) + 7/2$

$= 4(3 \times 11/2 - 2) + 7/2 = 4(33/2 - 2) + 7/2$

$= 4(29/2) + 7/2 = 58 + 7/2 = 116/2 + 7/2$

$= 123/2$

The answer on both sides is equal, hence the solution is proved. You must have noted in the above equations that, while solving a linear equation in one variable, we take into consideration the expression on both the sides one by one. This not only makes the concept clear but also helps us to reach to solutions faster.

Application of Linear Equations

Linear equations in one variable can be used to calculate the value of an integer when we are given two or more integers with one of them being unknown. Using the expression in an equation form, we can easily find the value of an unknown integer.

Linear equation helps us deduce integers in many word problems of arithmetic. Not only these expressions make the equations simple to understand but also help in reaching to conclusions easily and fast.

Example: In the last three months Mr. Sharma lost 5 and a half kg, gained 2 and one-fourth kg and then lost 3 and three fourth kg weight. If he now weighs 95 kg, then how much did Mr.Sharma weighs in beginning?

a. 100 kg

b. 102kg

c. 106.5 kg

d. 104kg

Solution: B. Let x be the Mr. Sharma's weight in beginning. According to given in instructions,

$$x - 5.5 + 2.5 - 3.75 = 95$$

$$\Rightarrow x = 95 + 5.5 + 3.75 - 2.25 = 95 + 7 = 102 \text{ kg.}$$

Linear Equations in Two Variables

An equation that can be put in the form ax + by + c = 0, where a, b and c are real numbers and a, b not equal to zero is called a linear equation in two variables namely x and y. The solution for such an equation is a pair of values, one for x and one for y which further makes the two sides of an equation equal.

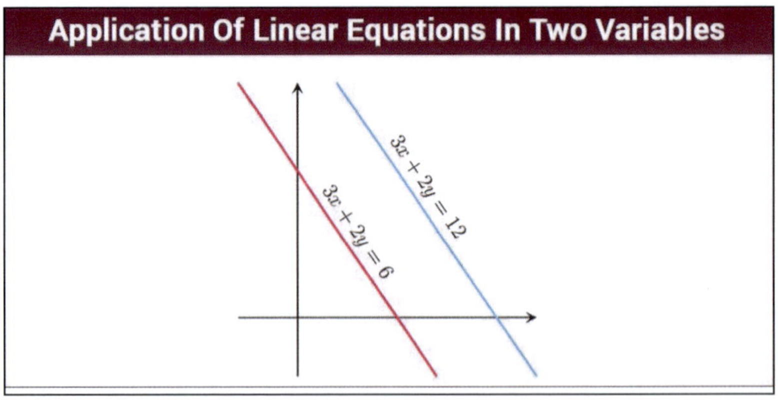

Let us take an example of a linear equation in two variables and understand the concept.

In order to find the solution of Linear equation in 2 variables, two equations should be known to us.

Consider for Example: $5x + 3y = 30$

The above equation has two variables namely x and y.

Graphically this equation can be represented by substituting the variables to zero.

The value of x when y = 0 is:

$$5x + 3(0) = 30$$

$$\Rightarrow x = 6$$

and the value of y when x = 0 is:

$$5(0) + 3y = 30$$

$$\Rightarrow y = 10$$

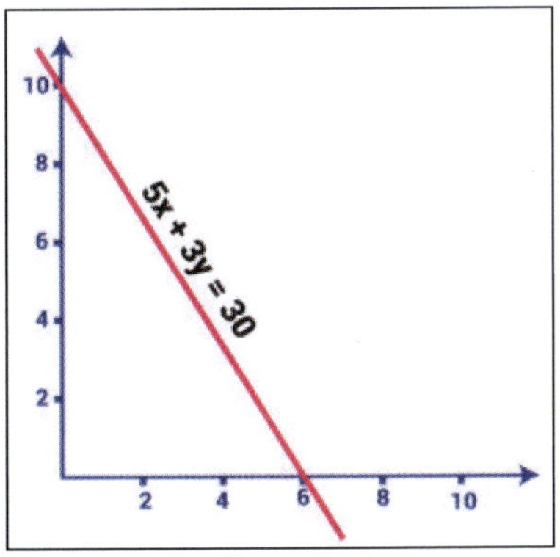

Quadratic Equation

A quadratic equation is an equation that could be written as:

$$ax^2 + bx + c = 0$$

when a 0.

There are three basic methods for solving quadratic equations: factoring, using the quadratic formula, and completing the square.

Factoring

To solve a quadratic equation by factoring:

1. Put all terms on one side of the equal sign, leaving zero on the other side;
2. Factor;
3. Set each factor equal to zero;
4. Solve each of these equations;
5. Check by inserting your answer in the original equation.

Examples:

1. Solve $x^2 - 6x = 16$

   Following the steps,

   $x^2 - 6x = 16$

becomes,

$x^2 - 6x - 16 = 0$

Factor,

$(x - 8)(x + 2) = 0$

Setting each factor to zero,

$x - 8 = 0 \text{ or } x + 2 = 0$

$x = 8 \qquad x = -2$

Then to check,

$8^2 - 6(8) = 16 \quad or \quad (-2)^2 - 6(-2) = 16$

$64 - 48 = 16 \qquad 4 + 12 = 16$

$16 = 16 \qquad 16 = 16$

Both values, 8 and −2, are solutions to the original equation.

2. Solve $y^2 = -6y - 5$

   Setting all terms equal to zero,

   $y^2 + 6y + 5 = 0$

   Factor,

   $(y + 5)(y + 1) = 0$

   Setting each factor to 0,

   $y + 5 = 0 \quad or \quad y + 1 = 0$

   $y = -5 \qquad y = -1$

   To check, $y^2 = -6y - 5$

   $(-5)^2 = -6(-5) - 5 \quad or \quad (-1)^2 = -6(-1) - 5$

   $25 = 30 - 5 \qquad 1 = 6 - 5$

   $25 = 25 \qquad 1 = 1$

A quadratic with a term missing is called an incomplete quadratic (as long as the ax^2 term isn't missing).

3. Solve $x^2 - 16 = 0$

   Factor,

   $(x + 4)(x - 4) = 0$

   $x + 4 = 0 \quad or \quad x - 4 = 0$

   $x = -4 \qquad x = 4$

To check, $x^2 - 16 = 0$

$$(-4)^2 + 16 = 0 \quad or \quad (4)^2 + 16 = 0$$
$$16 - 16 = 0 \qquad 16 + 16 = 0$$
$$0 = 0 \qquad 0 = 0$$

4. Solve $x^2 - 6x = 0$

$$x(x+6) = 0$$
$$x = 0 \quad or \quad x + 6 = 0$$
$$x = 0 \qquad x = -6$$

To check, $x^2 + 6x = 0$

$$(0)^2 + 6(0) = 0 \quad or \quad (-6)^2 + 6(-6) = 0$$
$$0 + 0 = 0 \qquad 36 + (-36) = 0$$
$$0 = 0 \qquad 0 = 0$$

5. Solve $2x^2 + 2x - 1 = x^2 + 6x - 5$.

First, simplify by putting all terms on one side and combining like terms.

$$2x^2 + 2x - 1 = x^2 + 6x - 5$$
$$\underline{-x^2 - 6x + 5 \quad -x^2 - 6x + 5}$$
$$x^2 - 4x + 4 = 0$$

Now, factor:

$$(x-2)(x-2) = 0$$
$$x - 2 = 0$$
$$x = 2$$

To check, $2x^2 + 2x - 1 = x^2 + 6x - 5$

$$2(2)^2 + 2(2) - 1 = (2)^2 + 6(2) - 5$$
$$8 + 4 - 1 = 4 + 12 - 5$$
$$11 = 11$$

The Quadratic Formula

Many quadratic equations cannot be solved by factoring. This is generally true when the roots, or answers, are not rational numbers. A second method of solving quadratic equations involves the use of the following formula:

$$x = \frac{-b \pm \sqrt{b^2 - 4ac}}{2a}$$

a, b, and c are taken from the quadratic equation written in its general form of:

$ax^2 + bx + c = 0$

Where a is the numeral that goes in front of x^2, b is the numeral that goes in front of x, and c is the numeral with no variable next to it (a.k.a., "the constant").

When using the quadratic formula, you should be aware of three possibilities. These three possibilities are distinguished by a part of the formula called the discriminant. The *discriminant* is the value under the radical sign, $b^2 - 4ac$. A quadratic equation with real numbers as coefficients can have the following:

1. Two different real roots if the discriminant $b^2 - 4ac$ is a positive number.
2. One real roots if the discriminant $b^2 - 4ac$ is equal to 0.
3. No real root if the discriminant $b^2 - 4ac$ is a negative number.

Examples:

1. Solve for x: $x^2 - 5x = -6$.

   Setting all terms equal to 0,

   $x^2 - 5x + 6 = 0$

Then substitute 1 (which is understood to be in front of the x^2), −5, and 6 for a, b, and c, respectively, in the quadratic formula and simplify.

$$x = \frac{-b \pm \sqrt{b^2 - 4ac}}{2a}$$

$$x = \frac{-(-5) \pm \sqrt{(-5)^2 - 4(1)(6)}}{2(1)}$$

$$x = \frac{5 \pm \sqrt{25 - 24}}{2}$$

$$x = \frac{5 \pm \sqrt{1}}{2}$$

$$x = \frac{5 \pm 1}{2}$$

$$x = \frac{5+1}{2} \quad or \quad x = \frac{5-1}{2}$$

$$x = \frac{6}{2} \quad or \quad x = \frac{4}{2}$$

$$x = 3 \quad or \quad x = 2$$

Because the discriminant $b^2 - 4ac$ is positive, you get two different real roots.

Example produces rational roots. In Example, the quadratic formula is used to solve an equation whose roots are not rational.

2. Solve for y: $y^2 = -2y + 2$.

   Setting all terms equal to 0,

   $y^2 + 2y - 2 = 0$

Then substitute 1, 2, and −2 for a, b, and c, respectively, in the quadratic formula and simplify.

$$y = \frac{-b \pm \sqrt{b^2 - 4ac}}{2a}$$
$$y = \frac{-(2) \pm \sqrt{(2)^2 - 4(1)(-2)}}{2(1)}$$
$$y = \frac{-2 \pm \sqrt{4+8}}{2}$$
$$y = \frac{-2 \pm \sqrt{12}}{2}$$
$$y = \frac{-2 \pm \sqrt{4}\sqrt{3}}{2}$$
$$y = \frac{-2 \pm 2\sqrt{3}}{2}$$
$$y = \frac{{}^{1}\not{2}\left(-1 \pm \sqrt{3}\right)}{\not{2}_{1}}$$
$$y = -1 + \sqrt{3} \text{ or } y = -1 - \sqrt{3}$$

Note that the two roots are irrational.

3. Solve for x: $x^2 + 2x + 1 = 0$.

Substituting in the quadratic formula,

$$x = \frac{-b \pm \sqrt{b^2 - 4ac}}{2a}$$
$$x = \frac{-(2) \pm \sqrt{(2)^2 - 4(1)(1)}}{2(1)}$$
$$x = \frac{-2 \pm \sqrt{4-4}}{2}$$

$$x = \frac{-2 \pm \sqrt{0}}{3}$$

$$x = \frac{-2}{2} = -1$$

Since the discriminant $b^2 - 4ac$ is 0, the equation has one root.

The quadratic formula can also be used to solve quadratic equations whose roots are imaginary numbers, that is, they have no solution in the real number system.

4. Solve for x: $x(x + 2) + 2 = 0$, or $x^2 + 2x + 2 = 0$.

   Substituting in the quadratic formula,

$$x = \frac{-b \pm \sqrt{b^2 - 4ac}}{2a}$$

$$x = \frac{-(2) \pm \sqrt{(2)^2 - 4(1)(2)}}{2(1)}$$

$$x = \frac{-2 \pm \sqrt{4-8}}{2}$$

$$x = \frac{-2 \pm \sqrt{-4}}{2}$$

Since the discriminant $b^2 - 4ac$ is negative, this equation has no solution in the real number system.

But if you were to express the solution using imaginary numbers, the solutions would be:

$$x = \frac{-2 \pm \sqrt{-4}}{2} = \frac{-2 \pm 2i}{2} = -1 \pm i \text{ or } -1+i, -1-i.$$

Completing the Square

A third method of solving quadratic equations that works with both real and imaginary roots is called completing the square.

- Put the equation into the form $ax^2 + bx = -c$.
- Make sure that $a = 1$ (if $a \neq 1$, multiply through the equation by $\frac{1}{a}$ before proceeding).
- Using the value of b from this new equation, add $\left(\frac{b}{2}\right)^2$ to both sides of the equation to form a perfect square on the left side of the equation.
- Find the square root of both sides of the equation.
- Solve the resulting equation.

5. Solve for x: $x^2 - 6x + 5 = 0$.

Arrange in the form of:

$$ax^2 + bx = -c$$
$$x^2 - 6x = -5$$

Because $a = 1$, add $\left(\frac{-6}{2}\right)^2$, or 9, to both sides to complete the square.

$$x^2 - 6x + 9 = -5 + 9$$
$$x^2 - 6x + 9 = 4$$
$$(x-3)^2 = 4$$

Take the square root of both sides.

$$x - 3 = \pm 2$$

Solve:

$$x - 3 = \pm 2$$
$$x - 3 = +2 \quad or \quad x - 3 = -2$$
$$+3 + 3 \qquad\qquad +3 + 3$$
$$x = 5 \quad or \qquad x = 1$$

6. Solve for y: $y^2 + 2y - 4 = 0$.

Arrange in the form of:

$$ay^2 + by = -c$$
$$y^2 + 2y = 4$$

Because $a = 1$, add $\left(\frac{2}{2}\right)^2$, or 1, to both sides to complete the square.

$$y^2 + 2y + 1 = 4 + 1$$
$$y^2 + 2y + 1 = 5$$
$$(y+1)^2 = 5$$

Take the square root of both sides.

$$y + 1 = \pm\sqrt{5}$$

Solve:

$$y + 1 = \pm\sqrt{5}$$
$$-1 \qquad -1$$

$y = -1 \pm \sqrt{5}$

$y = -1 + \sqrt{5} \ \ or \ \ y = -1 - \sqrt{5}$

7. Solve for x: $2x^2 + 3x + 2 = 0$.

   Arrange in the form of:

   $ax^2 + bx = -c$

   $2x^2 + 3x = -2$

Because a ≠ 1, multiply through the equation by:

$\frac{1}{2}.$

$x^2 + \frac{3}{2}x = -1$

Add $\left[\left(\frac{1}{2}\right)\left(\frac{3}{2}\right)\right]^2$ or $\frac{9}{16}$ to both sides.

$x^2 + \frac{3}{2}x + \frac{9}{16} = -1 + \frac{9}{16}$

$x^2 + \frac{3}{2}x + \frac{9}{16} = -\frac{7}{16}$

$\left(x + \frac{3}{4}\right)^2 = -\frac{7}{16}$

Take the square root of both sides.

$x + \frac{3}{4} = \pm\sqrt{\frac{-7}{16}}$

$x + \frac{3}{4} = \pm\frac{\sqrt{-7}}{\sqrt{16}}$

$x + \frac{3}{4} = \pm\frac{\sqrt{-7}}{4}$

$x = -\frac{3}{4} \pm \frac{\sqrt{-7}}{4}$

There is no solution in the real number system. It may interest you to know that the completing the square process for solving quadratic equations was used on the equation $ax^2 + bx + c = 0$ to derive the quadratic formula.

Cubic Equation

Equations of the third degree are called cubic equations. The general form of a cubic is:

$$x^3 + bx^2 + cx + d = 0.$$

Cubic equations have three possible values for *x*, at least one of which is real. There are several methods for solving cubic equations: factorization, rational root theorem, Descartes rule of signs, Vieta's root theorem, and Cardano's method.

Factoring Cubic Polynomials

A cubic polynomial is of the form:

$$p(x) = a_3 x^3 + a_2 x^2 + a_1 x + a_0.$$

The Fundamental Theorem of Algebra guarantees that if a_0, a_1, a_2, a_3 are all real numbers, then we can factor my polynomial into the form:

$$p(x) = a_3 (x - b_1)(x^2 + b_2 c + b_3).$$

Sometimes we can factor even further into the form:

$$p(x) = a_3 (x - c_1) (x - c_2)(x - c_3),$$

where c_1, c_2, c_3 are real numbers, but this is not always the case.

Factoring by Grouping

This is by far the nicest method of the two, but it only works in some cases. Consider the polynomial,

$$p(x) = x^3 - 4x^2 + 3x - 12.$$

We group the first two terms and the last two terms together:

$$p(x) = (x^3 - 4x^2) + (3x - 12)$$

and then we pull out the common factors:

$$p(x) = x^2 (x - 4) + 3(x - 4).$$

Notice now that these two terms now have x – 4 in common with each other; factor it out:

$$p(x) = (x - 4) (x^2 + 3).$$

$x^2 + 3$ is an irreducible quadratic, so it cannot factor into real terms. However, we can use the quadratic formula to solve for the roots.

Factoring using the Rational Root Theorem

This method works as long as the coefficients a_0, a_1, a_2, a_3 are all rational numbers. The Rational

Root Theorem says that the possible roots of a polynomial are the factors of the last term divided by the factors of the first term. In our case, since we are factoring the cubic polynomial above, the possible roots are:

$$\pm \frac{\text{factors of } a_0}{\text{factors of } a_3}$$

Example: List the possible roots of the following polynomials,

1. $p(x) = 4x^2 + 8x - 5x + 10$ The factors of 10 are ± 1, 2, 5, 10, and the factors of 4 are 1, 2, 4. Therefore the possible zeros of p(x) are:

$$\pm \frac{1,2,5,10}{1,2,4} = \pm 1, \frac{1}{2}, \frac{1}{4}, 2, 5, \frac{5}{2}, \frac{5}{4}, 10.$$

2. $p(x) = -x^3 + 9x^2 - 13x + 18$ The factors of 18 are ± 1, 2, 3, 6, 9 and the factors of −1 are ± 1. Therefore the possible zeros of p(x) are:

$$\pm \frac{1,2,3,6,9}{\pm 1} = \pm 1,2,3,6,9.$$

Now, let's put the theorem to action. Consider the polynomial,

$$p(x) = x^3 + 5x^2 - 2x - 24.$$

We cannot factor this polynomial by grouping, so we turn to the Rational Root Theorem. The factors of −24 are ± 1, 2, 3, 4, 6, 8, 12, 24, and the possible factors of 1 are ± 1. Therefore the possible zeros of p(x) are ±1, 2, 3, 4, 6, 8, 12, 24. We will then plug each of these numbers into p(x) until we find a root such that p = 0. Let's start with 1:

$$p(1) = 1 + 5 - 2 - 24 \neq 0$$

and so 1 is not a zero. Let's try -1:

$$p(-1) = -1 + 5 + 2 - 24 \neq 0,$$

and so -1 is not a zero. Let's try 2:

$$p(2) = 8 + 20 - 4 - 24 = 0,$$

and so 2 is a zero. This means that x − 2 is a factor of our polynomial. Now we use long division to divide p(x) by x − 2 to find out what the other possible zeros could be:

$$\frac{x^3 + 5x^2 - 2x - 24}{x - 2} = x^2 + 7x + 12.$$

This means that $p(x) = (x - 2)(x^2 + 7x + 12)$. The last term factors nicely:

$$p(x) = (x - 2)(x + 3)(x + 4),$$

and we are done.

Solution of Cubic Equations

We are going to find the exact solution of a general cubic equation:

$$ax^3 + bx^2 + cx + d = 0$$

To find the roots of Equation $ax^3 + bx^2 + cx + d = 0$, we first get rid of the quadratic term (x^2) by making the substitution:

$$x = y - \frac{b}{3a}$$

to obtain,

$$a\left(y - \frac{b}{3a}\right)^3 + b\left(y - \frac{b}{3a}\right)^2 + c\left(y - \frac{b}{3a}\right) + d = 0$$

Expanding previous equation above and simplifying, we obtain the following equation:

$$ay^3 + \left(c - \frac{b^2}{3a}\right)y + \left(d + \frac{2b^3}{27a^2} - \frac{bc}{3a}\right) = 0$$

Equation above is called the depressed cubic since the quadratic term is absent. Having the equation in this form makes it easier to solve for the roots of the cubic equation. First, convert the depressed cubic previous equation into the form:

$$y^3 + \frac{1}{a}\left(c - \frac{b^2}{3a}\right)y + \frac{1}{a}\left(d + \frac{2b^3}{27a^2} - \frac{bc}{3a}\right) = 0$$

$$y^3 + ey + f = 0$$

where,

$$e = \frac{1}{a}\left(c - \frac{b^2}{3a}\right)$$

$$f = \frac{1}{a}\left(d + \frac{2b^3}{27a^2} - \frac{bc}{3a}\right)$$

Now, reduce the above equation using Vieta's substitution:

$$y = z + \frac{s}{z}$$

For the time being, the constant s is undefined. Substituting into the depressed cubic equation

$$y^3 + ey + f = 0,$$

we get:

$$\left(z+\frac{s}{z}\right)^3+e\left(z+\frac{s}{z}\right)+f=0$$

Expanding out and multiplying both sides by z^3, we get:

$$z^6+(3s+e)z^4+fz^3+s(3s+e)z^2+s^3=0$$

Now, let $s=-\frac{e}{3}$ (s is no longer undefined) to simplify the equation into a tri-quadratic equation.

$$z^6+fz^3-\frac{e^3}{27}=0$$

By making one more substitution, $w = z^3$, we now have a general quadratic equation which can be solved using the quadratic formula.

$$w^2+fw-\frac{e^3}{27}=0$$

Once you obtain the solution to this quadratic equation, back substitute using the previous substitutions to obtain the roots to the general cubic equation.

$$w \rightarrow z \rightarrow y \rightarrow x$$

where we assumed,

$$w=z^3$$

$$y=z+\frac{s}{z}$$

$$s=-\frac{e}{3}$$

$$x=y-\frac{b}{3a}$$

You will get two roots for as Equation $w^2+fw-\frac{e^3}{27}=0$ is a quadratic equation. Using the above equation would then give you three roots for each of the two roots of z, hence giving you six root values for w, hence giving you six root z. But the six root values of z would give you six values of y; but three values of x will be identical to the other three. So one gets only three values of y, and hence three values of x. (Equation $x=y-\frac{b}{3a}$)

Examples:

1. Find the roots of the following cubic equation.

$$x^3-9x^2+36x-80=0$$

Solution: For the general form given by Equation $ax^3 + bx^2 + cx + d = 0$

$$ax^3 + bx^2 + cx + d = 0$$

We have,

$$a = 1,\ b = -9,\ c = 36,\ d = -80$$

in

$$x^3 - 9x^2 + 36x - 80 = 0$$

Equation $x^3 - 9x^2 + 36x - 80 = 0$ is reduced to:

$$y^3 + ey + f = 0$$

Where,

$$\begin{aligned} e &= \frac{1}{a}\left(c - \frac{b^2}{3a}\right) \\ &= \frac{1}{1}\left(36 - \frac{(-9)^2}{3(1)}\right) \\ &= 9 \end{aligned}$$

and

$$\begin{aligned} f &= \frac{1}{a}\left(d + \frac{2b^3}{27a^2} - \frac{bc}{3a}\right) \\ &= \frac{1}{1}\left(-80 + \frac{2(-9)^3}{27(1)^2} - \frac{(-9)(36)}{3(1)}\right) \\ &= -26 \end{aligned}$$

Giving,

$$y^3 + 9y - 26 = 0$$

For the general form given by Equation $y^3 + ey + f = 0$:

$$y^3 + ey + f = 0$$

we have,

e = 9 , f = -26

in Equation $y^3 + 9y - 26 = 0$.

From Equation $s=-\frac{e}{3}$

$$s=-\frac{e}{3}$$
$$=-\frac{9}{3}$$
$$=-3$$

From Equation $w^2+fw-\frac{e^3}{27}=0$

$$w^2+fw-\frac{e^3}{27}=0$$
$$w^2-26w-\frac{9^3}{27}=0$$
$$w^2-26w-27=0$$

where,

$$w=z^3$$

and

$$y=z+\frac{s}{z}$$
$$=z-\frac{3}{5}$$
$$w=\frac{-(-26)\pm\sqrt{(-26)^2-4(1)(-27)}}{2(1)}$$
$$=27,-1$$

The solution is:

$$w_1=27$$
$$w_2=-1$$

Since,

$$w=z^3$$
$$z^3=w$$

For $w = w_1$

$$z^3=w_1$$
$$=27$$
$$=27e^{i0}$$

Since,

$$w = z^3$$
$$re^{i\theta} = \left(ue^{i\alpha}\right)^3 = u^3 e^{3i\alpha}$$
$$r(\cos\theta + i\sin\theta) = u^3(\cos 3\alpha + i\sin 3\alpha)$$

resulting in;

$$r = u^3$$
$$\cos\theta = \cos 3\alpha$$
$$\sin\theta = \sin 3\alpha$$

Since, sinθ and cosθ are periodic of 2π,

$$3\alpha = \theta + 2\pi k$$
$$\alpha = \frac{\theta + 2\pi k}{3}$$

k will take the value of 0, 1 and 2 before repeating the same values of α.

So,

$$\alpha = \frac{\theta + 2\pi k}{3}, k = 0, 1, 2$$
$$\alpha_1 = \frac{\theta}{3}$$
$$\alpha_2 = \frac{(\theta + 2\pi)}{3}$$
$$\alpha_3 = \frac{(\theta + 4\pi)}{3}$$

So, roots of $w = z^3$ are:

$$z_1 = r^{\frac{1}{3}}\left(\cos\frac{\theta}{3} + i\sin\frac{\theta}{30}\right)$$
$$z_2 = r^{\frac{1}{3}}\left(\cos\frac{\theta + 2\pi}{3} + i\sin\frac{\theta + 2\pi}{3}\right)$$
$$z_3 = r^{\frac{1}{3}}\left(\cos\frac{\theta + 4\pi}{3} + i\sin\frac{\theta + 4\pi}{3}\right)$$

Gives,

$$z_1 = (27)^{1/3}\left(\cos\frac{0}{3} + i\sin\frac{0}{3}\right)$$
$$= 3$$

$$z_2 = (27)^{1/3}\left(\cos\frac{0+2\pi}{3} + i\sin\frac{0+2\pi}{3}\right)$$
$$= 3\left(\cos\frac{2\pi}{3} + i\sin\frac{2\pi}{3}\right)$$
$$= 3\left(-\frac{1}{2} + i\frac{\sqrt{3}}{2}\right)$$
$$= -\frac{3}{2} + i\frac{3\sqrt{3}}{2}$$
$$z_3 = (27)^{1/3}\left(\cos\frac{0+4\pi}{3} + i\sin\frac{0+4\pi}{3}\right)$$
$$= 3\left(\cos\frac{4\pi}{3} + i\sin\frac{0+4\pi}{3}\right)$$
$$= 3\left(-\frac{1}{2} - i\frac{\sqrt{3}}{2}\right)$$
$$= -\frac{3}{2} - i\frac{3\sqrt{3}}{2}$$

Since,

$$y = z - \frac{3}{z}$$
$$y_1 = z_1 - \frac{3}{z_1}$$
$$= 3 - \frac{3}{3}$$
$$= 2$$

$$y_2 = z_2 - \frac{3}{z_2}$$
$$= \left(-\frac{3}{2} + i\frac{3\sqrt{3}}{2}\right) - \frac{3}{\left(-\frac{3}{2} + i\frac{3\sqrt{3}}{2}\right)}$$
$$= -\frac{5+3i\sqrt{3}}{-1+i\sqrt{3}}$$
$$= -\frac{5+3i\sqrt{3}}{-1+i\sqrt{3}} \times \frac{-1-i\sqrt{3}}{-1-i\sqrt{3}}$$
$$= -1 + i2\sqrt{3}$$

$$\begin{aligned} y_3 &= z_3 - \frac{3}{z_3} \\ &= \left(-\frac{3}{2} - i\frac{3\sqrt{3}}{2}\right) - \frac{3}{\left(-\frac{3}{2} - i\frac{3\sqrt{3}}{2}\right)} \\ &= \frac{5 - i3\sqrt{3}}{1 + i\sqrt{3}} \\ &= \frac{5 - i3\sqrt{3}}{1 + i\sqrt{3}} \times \frac{1 + i\sqrt{3}}{1 - i\sqrt{3}} \\ &= -1 - i2\sqrt{3} \end{aligned}$$

Since,

$$\begin{aligned} x &= y + 3 \\ x_1 &= y_1 + 3 \\ &= 2 + 3 \\ &= 5 \\ x^2 &= y_2 + 3 \\ &= \left(-1 + i2\sqrt{3}\right) + 3 \\ &= 2 + i2\sqrt{3} \\ x_3 &= y_3 + 3 \\ &= \left(-1 - i2\sqrt{3}\right) + 3 \\ &= 2 - i2\sqrt{3} \end{aligned}$$

The roots of the original cubic equation:

$$x^3 - 9x^2 + 36x - 80 = 0$$

are x_1, x_2 , x_3 and , that is,

$$5, 2 + i2\sqrt{3},\ 2 - i2\sqrt{3}$$

Verifying,

$$(x - 5)\left(x - \left(2 + i2\sqrt{3}\right)\right)\left(x - \left(2 - i2\sqrt{3}\right)\right) = 0$$

Gives,

$$x^3 - 9x^2 + 36x - 80 = 0$$

Using,

$$w_2 = -1$$

would yield the same values of the three roots of the equation.

2. Find the roots of the following cubic equation:

$$x^3 - 0.03x^2 + 2.4\times10^{-6} = 0$$

Solution: For the general form:

$$ax^3 + bx^2 + cx + d = 0$$

$$a = 1,\ b = -0.03,\ c = 0,\ d = 2.4\times10^{-6}$$

Depress the cubic equation by letting (Equation $x = y - \frac{b}{3a}$)

$$\begin{aligned} x &= y - \frac{b}{3a} \\ &= y - \frac{(-0.03)}{3(1)} \\ &= y + 0.01 \end{aligned}$$

Substituting the above equation into the cubic equation and simplifying, we get:

$$y^3 - \left(3\times10^{-4}\right)y + \left(4\times10^{-7}\right) = 0$$

That gives $e = -3\times10^{-4}$ and $f = 4\times10^{-7}$ for equation $y^3 + ey + f = 0$, that is, $y^3 + ey + f = 0$

Now, solve the depressed cubic equation by using Vieta's substitution as:

$$y = z + \frac{s}{z}$$

to obtain,

$$z^6 + \left(3s - 3\times10^{-4}\right)z^4 + \left(4\times10^{-7}\right)z^3 + s\left(3s - 3\times10^{-4}\right)z^2 + s^3 = 0$$

Letting,

$$s = -\frac{e}{3} = -\frac{-3\times10^{-4}}{3} = 10^{-4}$$

we get the following tri-quadratic equation:

$$z^6 + \left(4\times10^{-7}\right)z^3 + 1\times10^{-12} = 0$$

Using the following conversion, w = z^3, we get a general quadratic equation:

$$w^2+\left(4\times10^{-7}\right)w+\left(1\times10^{-12}\right)=0$$

Using the quadratic equation, the solutions for w are:

$$w=\frac{-4\times10^{-7}\pm\sqrt{\left(4\times10^{-7}\right)^2-4(1)\left(1\times10^{-12}\right)}}{2(1)}$$

Giving,

$$w_1=-2\times10^{-7}+i\left(9.79795897113\times10^{-7}\right)$$
$$w_2=-2\times10^{-7}+i\left(9.79795897113\times10^{-7}\right)$$

Each solution w = z^3 of yields three values of z. The three values of z from w_1 are in rectangular form.

Since,

$$w = z^3$$

Then,

$$z=w^{\frac{1}{3}}$$

Let,

$$w=r\left(\cos\theta+i\sin\theta\right)=re^{i\theta}$$

Then,

$$z=u\left(\cos\alpha+i\sin\alpha\right)=ue^{i\alpha}$$

This gives,

$$w=z^3$$
$$re^{i\theta}=\left(ue^{i\alpha}\right)^3=u^3e^{3i\alpha}$$
$$r\left(\cos\theta+i\sin\theta\right)=u^3\left(\cos3\alpha+i\sin3\alpha\right)$$

resulting in,

$$r=u^3$$
$$\cos\theta=\cos3\alpha$$
$$\sin\theta=\sin3\alpha$$

Since sinθ and cosθ are periodic of 2π,

$$3\alpha = \theta + 2\pi k$$
$$\alpha = \frac{\theta + 2\pi k}{3}$$

k will take the value of 0, 1 and 2 before repeating the same values of α.

So,

$$\alpha = \frac{\theta + 2\pi k}{3}, k = 0,1,2$$
$$\alpha_1 = \frac{\theta}{3}$$
$$\alpha_2 = \frac{(\theta + 2\pi)}{3}$$
$$\alpha_3 = \frac{(\theta + 4\pi)}{3}$$

So, the roots of w = z³ are :

$$z_1 = r^{\frac{1}{3}}\left(\cos\frac{\theta}{3} + i\sin\frac{\theta}{3}\right)$$
$$z_2 = r^{\frac{1}{3}}\left(\cos\frac{\theta + 2\pi}{3} + i\sin\frac{\theta + 2\pi}{3}\right)$$
$$z_3 = r^{\frac{1}{3}}\left(\cos\frac{\theta + 4\pi}{3} + i\sin\frac{\theta + 4\pi}{3}\right)$$

So, for

$$w_1 = -2\times10^{-7} + i\left(9.79795897113\times10^{-7}\right)$$
$$r = \sqrt{\left(-2\times10^{-7}\right)^2 + \left(9.79795897113\times10^{-7}\right)^2}$$
$$= 1\times10^{-6}$$
$$\theta = \tan^{-1}\frac{9.79795897113\times10^{-7}}{-2\times10^{-7}}$$
$$= 772154248.1$$

(2nd quadrant because y (the numerator) is positive and x (the denominator) is negative.)

$$z_1 = \left(1\times10^{-6}\right)^{\frac{1}{3}}\left(\cos\frac{1.772154248}{3} + i\sin\frac{1.772154248}{3}\right)$$
$$= 0.008305409517 + i0.005569575635$$

$$z_2 = \left(1\times10^{-6}\right)^{\frac{1}{3}}\left(\cos\frac{1.772154248+2\pi}{3}+i\sin\frac{1.772154248+2\pi}{3}\right)$$
$$= -0.008976098746 + i0.004407907815$$
$$z_3 = \left(1\times10^{-6}\right)^{\frac{1}{3}}\left(\cos\frac{1.772154248+4\pi}{3}+i\sin\frac{1.772154248+4\pi}{3}\right)$$
$$= 0.0006706892313 - i0.009977483448$$

Compiling,

$$z_1 = 0.008305409518 + i0.005569575634$$
$$z_2 = -0.008976098746 + i0.004407907814$$
$$z_3 = 6.70689228525\times10^{-4} + i0.009977483448$$

Similarly, the three values of z from w_2 in rectangular form are:

$$z_4 = 0.008305409518 - i0.005569575634$$
$$z_5 = -0.008976098746 - i0.004407907814$$
$$z_6 = 6.70689228525\times10^{-4} + i0.009977483448$$

Using Vieta's substitution (Equation $y = z + \frac{s}{z}$),

$$y = z + \frac{s}{z}$$
$$y = z + \frac{\left(1\times10^{-4}\right)}{z}$$

we back substitute to find three values for y.

For example, choosing

$$z_1 = 0.008305409518 + i0.005569575634$$

Gives,

$$y_1 = 0.008305409518 + i0.005569575634 + \frac{1\times10^{-4}}{0.008305409518 + i0.005569575634}$$
$$= 0.008305409518 + i0.005569575634$$
$$+ \frac{1\times10^{-4}}{0.008305409078 + i0.00556957634}\times\frac{0.008305409518 - i0.00556957634}{0.008305409518 - i0.00556957634}$$
$$= 0.008305409518 + i0.005569575634 + \frac{1\times10^{-4}}{1\times10^{-4}}(0.008305409518 - i0.00556957634)$$
$$= 0.016610819036$$

The values of z_1, z_2 and z_3 give:

$$y_1 = 0.016610819036$$
$$y_2 = -0.01795219749$$
$$y_3 = 0.001341378457$$

respectively. The three other z values of z_4, z_5 and z_6 give the same values as y_1, y_2 and y_3, respectively.

Now, using the substitution of:

$$x = y + 0.01$$

The three roots of the given cubic equation are:

$$x_1 = 0.016610819036 + 0.01$$
$$= 0.026610819036$$
$$x_2 = -0.01795219749 + 0.01$$
$$= -0.00795219749$$
$$x_3 = 0.00134378457 + 0.01$$
$$= 0.011341378457$$

References

- Elementary-algebra, algebra: mathcaptain.com, Retrieved 13 February, 2019
- Solving-quadratic-equations, quadratic-equations, algebra-i, algebra, study-guides: cliffsnotes.com, Retrieved 17 March, 2019
- Real-number, science: britannica.com, Retrieved 23 June, 2019
- Real-numbers, maths: byjus.com, Retrieved 12 April, 2019
- Exponent, number, math: rapidtables.com, Retrieved 21 March , 2019
- How-to-add-and-subtract-with-powers, math, education: dummies.com, Retrieved 13 June, 2019
- Multiplying-exponents, exponent, number, math: rapidtables.com, Retrieved 24 March , 2019
- What-are-polynomials: mathbootcamps.com, Retrieved 13 July, 2019
- What-is-a-polynomial, stem: owlcation.com, Retrieved 21 July, 2019
- Zeroes-of-polynomial, polynomials, maths, guides: toppr.com, Retrieved 7 May, 2019
- Geometrical-representation-zeroes-polynomial, polynomials, maths, guides: toppr.com, Retrieved 18 April, 2019

5

Theorems in Algebra

There are a number of important theorems in algebra such as Cayley-Hamilton Theorem, Chebotarëv's theorem on roots of unity, the remainder theorem and the factor theorem, the fundamental theorem of ring homomorphisms and fundamental theorem of Galois Theory. The diverse applications of these theorems have been thoroughly discussed in this chapter.

Fundamental Theorem of Algebra

Let f be the previously defined polynomial. We first show that there exists at least one root of f in the complex numbers. With one root we can use an inductive argument to show that there are n roots, possibly with multiplicity greater than one. Let I be the interval [0, 1]. For positive real numbers r and R, let $\gamma: I \rightarrow \mathbb{C}$ and $\Gamma: I \rightarrow \mathbb{C}$ be paths in the complex plane defined by $\gamma(x) = re^{2\pi ix}$ and $\Gamma(x) = Re^{2\pi ix}$. These paths are circles of radius r and R. They are homotopic in C, so their images $f \circ \gamma$ and $f \circ \Gamma$ are homotopic because polynomials are continuous. To show that there is at least one root, we show that there are values of r and R such that $f \circ \gamma$ and $f \circ \Gamma$ are not homotopic in $\mathbb{C} \setminus \{0\}$. This will be a contradiction to what we have just shown, showing that 0 must lie in the image of f. Let $g_1 : I \rightarrow \mathbb{C}$ and $g_2 : I \rightarrow \mathbb{C}$ be defined as $g_1(x) = 1$ and $g_2(x) = e^{2\pi inx}$. We show that there exist values of r and R such that both $f \circ \gamma \simeq a_0 g_1$ and $f \circ \Gamma \simeq a_0 g_2$, and, moreover, that there exist homotopies between them that are nowhere zero. We then show that $a_0 g_1$ and $a_0 g_2$ are not homotopic in $\mathbb{C} \setminus \{0\}$, which will give our desired contradiction.

The existence of the first homotopy follows from the continuity of f. Note that $f(0) = a_0$. Then, there exists δ > 0 such that if $|z| < \delta$, then

$$|f(z) - a_0| < \left|\frac{a_0}{2}\right|.$$

Suppose that $r < \delta$. If z is in the path

$$|z| = r. \text{Then} |f(z) - a_0| < \left|\frac{a_0}{2}\right|$$

for $z \in \gamma$. By the reverse triangle inequality

$$\big||f(z)| - |a_0|\big| \leq |f(z) - a_0| < \left|\frac{a_0}{2}\right|,$$

Then,

$$-\left|\frac{a_0}{2}\right| < \left\|f(z)\right| - \left|a_0\right| < \left\|\frac{a_0}{2}\right|,$$

and, adding $|a_0|$ to both sides, $\left|\frac{a_0}{2}\right| < |f(z)| < \left|\frac{3a_0}{2}\right|$. Consider the homotopy between $f \circ \gamma$ and $a_0 g_1, h : I \times I \to \mathbb{C}$, defined as $h(x,t) = (1-t)(f \circ \gamma)(x) + a_0 t$. If $r < \delta$, then, $|h(x,t)| > 0$, as shown below:

$$\begin{aligned}|h(x,t)| &= (1-t)(f \circ \gamma)(x) + a_0 t\,| \\ &= \left|(f \circ \gamma)(x) - t\left((f \circ \gamma)(x) - a_0\right)\right| \\ &\geq \left|(f \circ \gamma)(x)\right| - t\left|(f \circ \gamma)(x) - a_0\right| \\ &\geq \left|\frac{a_0}{2}\right| - t\left|\frac{a_0}{2}\right| \\ &\geq 0\end{aligned}$$

Thus, with r chosen small enough, $f \circ \gamma \simeq a_0 g_1$ with a nowhere zero homotopy. To show that $f \circ \Gamma \simeq a_0 g_2$ with a nowhere zero homotopy, we first show that $f \circ \Gamma \simeq R^n g_2$ with a nowhere zero homotopy for large enough $|R|$. Then, we show that $R^n g_2 \simeq a_0 g_2$ with a nowhere zero homotopy.

For the first part, we make use of the fact that polynomials eventually behave like their leading term. We can pick some real M such that if $|z| > M$, then $|z|^{n-j} > 2n|a_j|$ where a_j is a coefficient in f and $0 \leq j < n$, Then, rearranging, $\left|\frac{z^n}{2n}\right| > \left|a_j z^j\right|$. By picking M large enough, we make this inequality true for all $0 \leq j < n$, Then, adding the ineqalities;

$$\left|\frac{z^n}{2}\right| > \left|a_n - 1z^{n-1}\right| + \cdots + |a_1 z| + |a_0| \geq \left|a_{n-1} z^{n-1} + \cdots + a_1 z + a_0\right|.$$

Then, with $z > M$,

$$\begin{aligned}|f(z)| &= \left|z^n + a_{n-1} z^{n-1} + \cdots + a_0\right| \\ &\geq |z^n| - \left|a_{n-1} z^{n-1} + \cdots + a_0\right| \\ &= \left|\frac{z^n}{2}\right| + \left(\left|\frac{z^n}{2}\right| - \left|a_{n-1} z^{n-1} + \cdots + a_0\right|\right) \\ &> \left|\frac{z^n}{2}\right|\end{aligned}$$

Let $|R| > M$, and consider the homotopy $h : I \times I \to \mathbb{C}$ defined by,

$$h(x,t) = (1-t)(f \circ \Gamma)(x) + t\left(R^n e^{2\pi inx}\right).$$

$$\begin{aligned}|h(x,t)| &= \left|(f \circ \Gamma)(x) - t\left((f \circ \Gamma)(x) - R^n e^{2\pi inx}\right)\right| \\ &= \left|\left(R^n e^{2\pi inx} + \cdots + a_0\right)\right. \\ &\quad \left. - t\left(a_{n-1}R^{n-1}e^{2\pi i(n-1)x} + \cdots + a_1 R e^{2\pi ix} + a_0\right)\right| \\ &> \left|\frac{R^n}{2}\right| - \left|a_{n-1}R^{n-1}e^{2\pi i(n-1)}x + \cdots + a_1 R e^{2\pi ix} + a_0\right| \\ &> 0\end{aligned}$$

We now show that $R^n g_2 \simeq a_0 g_2$ with a nowhere zero homotopy. Let $R^n = R^* e^{i\theta_1}$, and let $a_0 = a^* e^{i\theta_2}$, where R^* and a^* are positive real numbers. Consider the homotopy $h : I \times I \to \mathbb{C}$ defined by

$$h(x,t) = \left((1-t)R^* + ta^*\right)e^{i((1-t)\theta_1 + t\theta_2)}e^{2\pi inx}.$$

The function $(1-t)R^* + ta^*$ has a zero at;

$$t = \frac{R^*}{R^* - a^*}.$$

If $R^* = a^*$, then the function is constant with no zero. If $R^* < a^*$, then the zero is negative. If $R^* > a^*$, then the zero is greater than 1. In every case, $t \notin I$ when the function is zero. Thus, the homotopy h is never 0, and, by running each homotopy at twice speed; we have that $f \circ \Gamma \simeq a_0 g_2$ with a nowhere zero homotopy.

We now consider the fundamental group $\pi_1(\mathbb{C} \setminus \{0\}, a_0)$, and we show that $a_0 g_1$ and $a_0 g_2$ belong to different elements of the group. We use the fact that the complex plane is a covering space of $\mathbb{C} \setminus \{0\}$ and that the exponential function is a covering map. Also, the paths $a_0 g_1$ and $a_0 g_2$ have unique lifting to paths in $\mathbb{C}$ beginning at a_0. We show that the endpoints of these paths are different. This is sufficient to show that they belong to different equivalence classes in the fundamental group.

Consider the covering map $p : \mathbb{C} \to \mathbb{C} \setminus \{0\}$ defined as $p(z) = e^{iz}$. Let e_0 be the point in $\mathbb{C}$ such that $a_0 = e^{ie_0}$. We consider liftings of $a_0 g_1$ and $a_0 g_2$ that begin at a_0 in $\mathbb{C}$. They are $\widetilde{a_0 g_1} : I \to \mathbb{C}$, defined as

$$\widetilde{a_0 g_1}(x) \to e_0, \text{ and } a_0 g_2 : I \to \mathbb{C},$$

defined as

$$\widetilde{a_0 g_1}(x) \to 2\pi nx + e_0.$$

As a check,

$$\left(p\circ\widetilde{a_0 g_1}\right)(x)=e^{ie_0}=a_0=a_0 g_1(x),$$

that,

$$\left(p\circ\widetilde{a_0 g_1}\right)(x)=e^{i(2\pi nx+e_0)}=a_0 e^{2\pi inx}=a_0 g_2(x),$$

as required. Also, note that $\widetilde{a_0 g_1}(1)=e_0$ and that $\widetilde{a_0 g_1}(1)=2\pi n+e_0$. Because $n\geq 1$, the liftings have different ending points, so the paths $a_0 g_1$ and $a_0 g_2$ are not homotopic in $\mathbb{C}\setminus\{0\}$. It follows from the transitivity of homotopy relations that $f\circ\gamma$ and $f\circ\Gamma$ are also not homotopic in $\mathbb{C}\setminus\{0\}$. This is a contradiction of the fact that they are homotopic. Thus, the image of f must contain 0, so f has a root.

The existence of n roots follows from an inductive argument once it is known that at least one root exists. Let,

$$P(z)=z^n+a_{n-1}z^{n-1}+\cdots+a_0$$

be a polynomial of degree n. If z_0 is a root of P then $P(z)=(z-z_0Q(z))$, whene Q is a polynomial of degree $n-1$. Because Q also has a root, we can continue expanding in this manner until P is written as the product of n linear factors. Thus, P has a total of n roots. This completes our first proof of the fundamental theorem of algebra.

A Similar Proof using the Language of Complex Analysis

We now present a proof of the fundamental theorem of algebra that is similar to the above but written in the language of complex analysis. We show that for a large enough circle centered at the origin, the image of the circle will wrap around the origin n times, but, under the assumption that the image of the polynomial lies in $\mathbb{C}\setminus\{0\}$, it does not wrap around the origin at all.

Consider a circle $\Gamma=Re^{2\pi ix}$, with $x\in I$ and R chosen large enough that,

$$R^n e^{2\pi ixn}|>|a_{n-1}R^{(n-1)}e^{2\pi ix(n-1)}+\cdots+a_0|.$$

Then, using Rouche's Theorem,

$$\begin{aligned}\int_{f(\Gamma)}\frac{dz}{z}&=\int_{\Gamma}\frac{f'(z)}{f(z)}dz\\&=\int_{\Gamma}\frac{(z^n)'}{z^n}dz\\&=\int_{\Gamma}\frac{n}{z}dz\\&=\int_0^1\frac{n}{Re^{2\pi ix}}2\pi i\,Re^{2\pi ix}dx\end{aligned}$$

$$= \int_0^1 2\pi i n dx$$
$$= 2\pi i n$$

We now show that under the assumption that f is never 0, the integral must be 0. Polynomials are holomorphic functions, and the inverses of holomorphic functions are holomorphic wherever the function is non zero. If the image of f is contained in $\mathbb{C}\setminus\{0\}$, then $\frac{1}{f}$ is holomorphic everywhere. Also, f' is a polynomial of degree $n-1$, so it is holomorphic everywhere. Then, $\frac{f'}{f}$ is holomorphic everywhere. By Cauchy's integral theorem, the integral of $\frac{f'}{f}$ over a closed path is 0. Then, this is a contradiction, so the image of f must contain 0. Then, there is at least one zero of f. The existence of n zeros, with possible multiplicity, follows by induction as in the previous proof.

A Proof using the Maximum Modulus Principle

We now provide a proof of the fundamental theorem of algebra that makes use of the maximum modulus principle, i.e., the modulus of a holomorphic function on a connected, open set has no local maximum unless the function itself is constant. Consider the disk of radius R centered at the origin such that

$$\left|f\left(Re^{i\theta}\right)\right| > \frac{1}{2}\left|Re^{i\theta}\right|^n$$

and

$$\frac{1}{2}\left|Re^{i\theta}\right|^n > |a_0|.$$

There exists an R that satisfies these conditions, as shown in the first proof. Then, for

$$z \geq |R|, |f(z)| > |a_0|.$$

The function $|f|$ is continuous and the disk is compact, so, by the extreme value theorem, |f| has a minimum on the disk. Call it a, and let α be a point such that $f(\alpha) = a$. For any z on the boundary of the disk, $|f(z)| > |a_0| \geq a$. Thus, the minimum is not achieved on the boundary. It is achieved in the open disk of radius R centered at the origin. Suppose that $a \neq 0$. Then, f is never 0, so the function $\frac{1}{f}$ is holomorphic. This allows us to apply the maximum modulus principle to it. The open disk of radius R is an open connected subset $\mathbb{C}$, and $\frac{1}{a}$ is a local maximum of the function $\frac{1}{f}$. The maximum modulus principle then implies that $\left|\frac{1}{f}\right|$ is constant. However, $|f|$ is then constant, but this is a contradiction because polynomials are not constant. Thus, $a=0$, and α is a root of f. The existence of n roots follows as it does in the first proof.

A Proof using Liouville's Theorem

Liouville's Theorem, i.e. that a bounded, entire function is constant, provides a proof of the fundamental theorem of algebra that is very similar to the proof using the maximum modulus principle. Consider the same disk of radius R used in the previous proof. As above, there exists some α on the disk such that $|f(a)|$ is a minimum on the disk. We suppose again that $f(\alpha) \neq 0$. For any z such that

$$|z| \geq R, |f(z)| > |f(\alpha)|, \text{ so } \left|\frac{1}{f(\alpha)}\right| > \left|\frac{1}{f(z)}\right|. \text{ Then, } \left|\frac{1}{f(\alpha)}\right|$$

is a maximum of $\left|\frac{1}{f}\right|$ over the whole complex plane. As above, $\left|\frac{1}{f}\right|$ is holomorphic on all of $\mathbb{C}$.Then, by Liousville's Theorem, $\left|\frac{1}{f}\right|$ is constant, so $|f|$ is constant. This is again a contradiction, so $f(\alpha) = 0$.

There are again n roots, following from an induction argument.

Fundamental Theorem of Linear Algebra

An M by N matrix $\mathrm{A} \in R^{M \times N}$ can be expressed in terms of either its c_j or its $N\ M - D$ row vectors r_i^T :

$$\mathrm{A} = \begin{bmatrix} a_{11} & \cdots & a_{1N} \\ \vdots & \ddots & \vdots \\ a_{M1} & \cdots & a_{MN} \end{bmatrix} = [c_1, \cdots, c_N] = \begin{bmatrix} r_1^T \\ \vdots \\ r_M^T \end{bmatrix}$$

$$\mathrm{A}^T = \begin{bmatrix} a_{11} & \cdots & a_{M1} \\ \vdots & \ddots & \vdots \\ a_{1N} & \cdots & a_{MN} \end{bmatrix} = [r_1, \cdots, R_M] = \begin{bmatrix} c_1^T \\ \vdots \\ c_N^T \end{bmatrix}$$

Where r_i and c_j are respectively the ith row vector and j^th^ column vector (note that throughout the text, all vectors are assumed to be vertical):

$$c_j = \begin{bmatrix} a_{1j} \\ \vdots \\ a_{Mj} \end{bmatrix}, \quad (j = 1, \cdots, N)$$

$$\mathrm{r}_i = \begin{bmatrix} a_{i1} \\ \vdots \\ a_{iN} \end{bmatrix} \quad \text{i.e. } r_i^T = [a_{i1}, \cdots, a_{iN}] \ (i = 1, \cdots, M)$$

The M by N matrix $\underline{\mathrm{A}}$ in the linear function (a transformation) f(x) = Ax maps anN-D vector in the domain of the function, $\mathrm{x} \in R^N$, to an M-D vector in the codomain of the function, $\mathrm{Ax} \in R^M$.

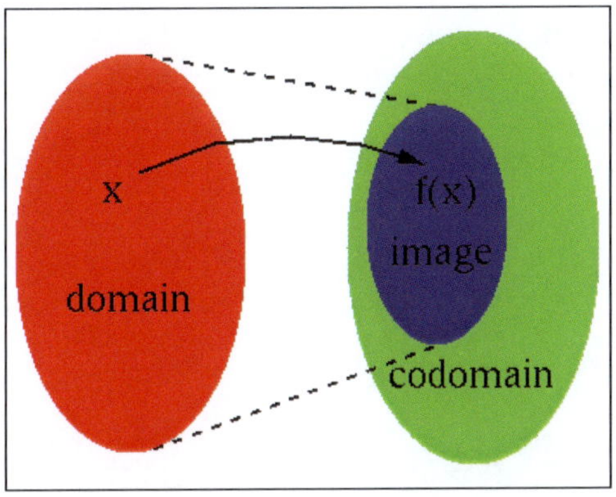

The fundamental theorem of linear algebra concerns the following four fundamental subspaces associated with any $M \times N$ matrix A with rank $R = rank(A) \leq \min(M, N)$, there are R independent columns and rows.

1. The column space of A is a space spanned by its N M–D column vectors ($R \leq N$ of which are independent):

   $$C(A) = span(c_1, \cdots, c_N) \subseteq R^M$$

   which is an R-D subspace of R^M composed of all possible linear combinations of its N column vectors:

   $$x_1 c_1 + \cdots + x_N c_N = [c_1, \cdots, c_N] \begin{bmatrix} x_1 \\ \vdots \\ x_N \end{bmatrix} = \text{Ax}, \text{ (for any x)}$$

   The column space $C(A)$ is the image (or range) of the linear transformation,

   $$\text{f(A)} = \text{Ax from } \overline{R^N} \text{ to } R^M.$$

   The equation system $\text{Ax} = \text{b}$ is solvable if and only $\text{b} \in C(\text{A})$. he dimension of the column space is the rank of A, $dim C(\text{A}) = rank(\text{A}) = R$.

2. The row space of A is a space spanned by its M N–D row vectors ($R \leq M$ of which are independent):

   $$R(\text{A}) = span(\text{r}_1, \cdots, \text{r}_M) \subseteq R^N$$

   which is an R-D subspace of R^N composed of all possible linear combinations of its M row vectors:

   $$x_1 \text{r}_1 + \cdots + x_M r_N = [\text{r}_1, \cdots, \text{r}_M \begin{bmatrix} x_1 \\ \vdots \\ x_M \end{bmatrix} = \text{A}^T \text{x}, \text{ (for any x)}$$

As the rows and columns in $\underline{\mathrm{A}}$ are respectively the columns and rows in A^T, the row space of $\underline{\mathrm{A}}$ is the column space of A^T, and the column space of $\underline{\mathrm{A}}$ is the row space of A^T:

$$R(\mathrm{A}) = C(A^T), \quad \text{and } C(\mathrm{A}) = R(\mathrm{A}^T)$$

The number of linearly independent row vectors is the dimension of the row space $dim\, R(\mathrm{A})$, which is the rank $R = rank(\mathrm{A})$,i.e.,

$$dim\, C(\mathrm{A}) = dim\, R(\mathrm{A}) = rank(\mathrm{A}) = R$$

3. The nullspace (kernel) of A, denoted by $N(\mathrm{A})$, is the set of all vectors $\underline{\mathrm{x}}$ that satisfy the homogeneous equation:

$$\mathrm{Ax} = \begin{bmatrix} r_1^T \\ \vdots \\ r_M^T \end{bmatrix} \mathrm{x} = 0, \quad \text{i.e.} \quad \mathrm{r}_i^T \mathrm{x} = 0, \; (i = 1, \cdots, M)$$

i.e.,

$$N(\mathrm{A}) = \left\{ \mathrm{x} \in R^N \mid \mathrm{Ax} = 0 \right\} \subseteq R^N$$

In particular, when $\mathrm{x} = 0$, we get $\mathrm{Ax} = 0$, i.e., the origin $0 \in N(\mathrm{A})$ must be in the null-space.

As all $x \in N(\mathrm{A})$ re orthogonal to $\mathrm{r}_i \in R(\mathrm{A}), N(\mathrm{A})$ is orthogonal to the row space $R(\mathrm{A}) = C(\mathrm{A}^T)$:

$$N(\mathrm{A}) \perp R(\mathrm{A}), \qquad N(\mathrm{A}) \perp C(\mathrm{A}^T)$$

The dimension of the nullspace is called the nullity of A: $dim\, N(\mathrm{A}) = nullity(\mathrm{A})$. Also, according to the rank-nullity theorem,

$$\mathrm{rank}(\mathrm{A}) + nullity(\mathrm{A}) = dim\, C(\mathrm{A}) + dim\, N(\mathrm{A}) = N$$

we see that $C(\mathrm{A})$ and $N(\mathrm{A})$ are mutually exclusive and complementary:

$$R(\mathrm{A}) \cap N(\mathrm{A}) = \theta, \quad R(\mathrm{A}) \perp N(\mathrm{A}), \quad R(\mathrm{A}) \oplus N(\mathrm{A}) = R^N$$

They are orthogonal complement of each other, denoted by:

$$N(\mathrm{A}) = R(\mathrm{A})^{\perp}, \quad R(\mathrm{A}) = N(\mathrm{A})^{\perp}$$

4. The nullspace of A^T (also called left nullspace of A), denoted by $N(\mathrm{A}^T)$, is the set of all vectors y that satisfy the homogeneous equation:

$$\mathrm{A}^T \mathrm{y} = [\mathrm{c}_1, \cdots, \mathrm{c}_N]^T \mathrm{y} = \begin{bmatrix} \mathrm{c}_1^T \\ \vdots \\ \mathrm{c}_N^T \end{bmatrix} \mathrm{y} = 0, \quad \text{i.e.} \quad \mathrm{c}_j^T \mathrm{y} = 0, \; (j = 1, \cdots, N)$$

i.e.,

$$N(\mathrm{A}^T) = \{\mathrm{y} \in R^M \mid \mathrm{A}^T \mathrm{y} = 0\} \subseteq \mathbb{R}^M$$

As all $\mathrm{y} \in N(\mathrm{A}^T)$ are orthogonal to

$$\mathrm{c}_i \in C(\mathrm{A}),\ N(\mathrm{A}^T)$$

is orthogonal to the column space $C(\mathrm{A}) = R(\mathrm{A}^T)$:

$$N(\mathrm{A}^T) \perp C(\mathrm{A}), \qquad N(\mathrm{A}^T) \perp R(\mathrm{A}^T)$$

Also, for $C(\mathrm{A}) = R(\mathrm{A}^T)$ and $N(\mathrm{A}^T)$ we have:

$$C(\mathrm{A}) \cap N(\mathrm{A}^T) = \emptyset, \quad C(\mathrm{A}) \perp N(\mathrm{A}^T), \quad C(\mathrm{A}) \oplus N(\mathrm{A}^T) = R^M,$$

$$N(\mathrm{A}^T) = C(\mathrm{A})^{\perp}, \ C(\mathrm{A}) = N(\mathrm{A}^T)^{\perp}$$

The bases of these four subspaces can be found by converting A into the reduced row echelon form (rref). by the elementary row operations (row reduction).

Example:

$$\mathrm{A} = \begin{bmatrix} 1 & 2 & 5 \\ 2 & 3 & 8 \\ 3 & 1 & 5 \end{bmatrix}$$

By elementary row operations, we can convert A into the rref form:

$$\mathrm{A} = \begin{bmatrix} 1 & 2 & 5 \\ 2 & 3 & 8 \\ 3 & 1 & 5 \end{bmatrix} \to \begin{bmatrix} 1 & 2 & 5 \\ 0 & -1 & -2 \\ 0 & -5 & -10 \end{bmatrix} \to \begin{bmatrix} 1 & 2 & 5 \\ 0 & 1 & 2 \\ 0 & 1 & 2 \end{bmatrix} \to \begin{bmatrix} 1 & 2 & 5 \\ 0 & 1 & 2 \\ 0 & 0 & 0 \end{bmatrix} \to \begin{bmatrix} 1 & 0 & 1 \\ 0 & 1 & 2 \\ 0 & 0 & 0 \end{bmatrix}$$

We see that $\mathrm{c}_3 = \mathrm{c}_1 + 2\mathrm{c}_2$. Alternatively, if we define another matrix A' by switching the 2nd and 3rd columns of A, we have;

$$\mathrm{A}' = \begin{bmatrix} 1 & 5 & 2 \\ 2 & 8 & 3 \\ 3 & 5 & 1 \end{bmatrix} \to \begin{bmatrix} 1 & 5 & 2 \\ 0 & -2 & -1 \\ 0 & -10 & -5 \end{bmatrix} \to \begin{bmatrix} 1 & 5 & 2 \\ 0 & 2 & 1 \\ 0 & 2 & 1 \end{bmatrix} \to \begin{bmatrix} 1 & 5 & 2 \\ 0 & 2 & 1 \\ 0 & 2 & 1 \end{bmatrix} \to \begin{bmatrix} 1 & 5 & 2 \\ 0 & 1 & 0.5 \\ 0 & 0 & 0 \end{bmatrix} \to \begin{bmatrix} 1 & 0 & -0.5 \\ 0 & 1 & 0.5 \\ 0 & 0 & 0 \end{bmatrix}$$

Now, we have $\mathrm{c}_3 = -0.5\mathrm{c}_1 + 0.5\mathrm{c}_2 = (\mathrm{c}_2 - \mathrm{c}_1)/2$. We see that in either case, the column space is a 2-D plane spanned by any two of the three column vectors.

The $\mathrm{R} = \mathrm{rank(A)}$ columns in the rref matrix that contain a single 1, called a pivot element, are the pivot columns, and the R corresponding rows each containing a pivot are the pivot rows, which span the R-D row space $\mathrm{R(A)}$. owever, the columns of A are changed by the row reduction operations and the pivot columns of the rref form do not span the column space $\mathrm{C(A)}$.

The number of pivots in the rref form is the rank of A, the number of independent rows or columns in A, i.e., the dimension of both $C(A)$ and $R(A)$.

$$dim\, C(A) = dim\, R(A) = rank(A)$$

The R pivot rows correspond to the independent equations in the system $Ax = b$, while the remaining $M - R$ non-pivot rows containing all zeros are not independent. They don't play any role in the equation system $Ax = b$, as they map any x to 0.

Example:

Solve the following homogeneous equation:

$$Ax = \begin{bmatrix} 1 & 2 & 5 \\ 2 & 3 & 8 \\ 3 & 1 & 5 \end{bmatrix} \begin{bmatrix} x_1 \\ x_2 \\ x_3 \end{bmatrix} = \begin{bmatrix} 0 \\ 0 \\ 0 \end{bmatrix} = 0$$

Applying the elementary row operations, we get:

$$\begin{bmatrix} 1 & 2 & 5 \\ 2 & 3 & 8 \\ 3 & 1 & 5 \end{bmatrix} \begin{bmatrix} x_1 \\ x_2 \\ x_3 \end{bmatrix} \rightarrow \begin{bmatrix} 1 & 2 & 5 \\ 0 & -1 & -5 \\ 0 & -5 & -10 \end{bmatrix} \begin{bmatrix} x_1 \\ x_2 \\ x_3 \end{bmatrix} \rightarrow \begin{bmatrix} 1 & 2 & 5 \\ 0 & 1 & 2 \\ 0 & 0 & 0 \end{bmatrix} \begin{bmatrix} x_1 \\ x_2 \\ x_3 \end{bmatrix} \rightarrow \begin{bmatrix} 1 & 0 & 1 \\ 0 & 1 & 2 \\ 0 & 0 & 0 \end{bmatrix} \begin{bmatrix} x_1 \\ x_2 \\ x_3 \end{bmatrix} = \begin{bmatrix} 0 \\ 0 \\ 0 \end{bmatrix}$$

In this rref form, there are only two pivot columns and pivot rows. The last row is all zeros, indicating that is singular ($R = 2 < M = N = 3$), i.e. x_3 is a free variable which can take any value. We have:

$$x_1 + x_3 = 0, \;\; x_2 + 2x_3 = 0$$

If we let then we get $x_1 = -1, x_2 = -2$, and a special solution $\mathbf{x}_n = [-1, -2, 1]^T$, which is in the nullspace $N(A)$. However, the free variable x_3 can take any value c, the complete solution, i.e., the entire nullspace is:

$$N(A) = c\,x_n = c\begin{bmatrix} -1 \\ -2 \\ 1 \end{bmatrix}$$

Example:

Solve the following non-homogeneous equation:

$$Ax = \begin{bmatrix} 1 & 2 & 5 \\ 2 & 3 & 8 \\ 3 & 1 & 5 \end{bmatrix} \begin{bmatrix} x_1 \\ x_2 \\ x_3 \end{bmatrix} = \begin{bmatrix} 19 \\ 31 \\ 22 \end{bmatrix} = b$$

We use Gauss-Jordan elimination to solve this system by applying a sequence of elementary row operations:

$$[\mathrm{A}|\mathrm{b}] = \begin{bmatrix} 1 & 2 & 5 & 19 \\ 2 & 3 & 8 & 31 \\ 3 & 1 & 5 & 22 \end{bmatrix} \rightarrow \begin{bmatrix} 1 & 2 & 5 & 19 \\ 0 & -1 & -2 & -7 \\ 0 & -5 & -10 & -35 \end{bmatrix} \rightarrow \begin{bmatrix} 1 & 2 & 5 & 19 \\ 0 & 1 & 2 & 7 \\ 0 & 0 & 0 & 0 \end{bmatrix} \rightarrow \begin{bmatrix} 1 & 0 & 1 & 5 \\ 0 & 1 & 2 & 7 \\ 0 & 0 & 0 & 0 \end{bmatrix}$$

There are only two independent columns (or rows) in the rref form, i.e., the rank is

$R = 2 < M = N = 3$, indicating that A it is singular, and its inverse A^{-1} does not exist, we cannot convert A into an identity matrix I Instead we just have the following system with a coefficient matrix in the rref form, which can also be expressed in block matrix form:

$$\begin{bmatrix} 1 & 0 & 1 \\ 0 & 1 & 2 \\ 0 & 0 & 0 \end{bmatrix} \begin{bmatrix} x_1 \\ x_2 \\ x_3 \end{bmatrix} = \begin{bmatrix} 5 \\ 7 \\ 0 \end{bmatrix} \quad \text{or} \quad \begin{bmatrix} \mathrm{I} & \mathrm{F} \\ 0 & 0 \end{bmatrix} \begin{bmatrix} \mathrm{X}_{privot} \\ \mathrm{X}_{free} \end{bmatrix} = \begin{bmatrix} \mathrm{b}_1 \\ 0 \end{bmatrix}$$

where,

$$\mathrm{I} = \begin{bmatrix} 1 & 0 \\ 0 & 1 \end{bmatrix}, \quad \mathrm{F} = \begin{bmatrix} 1 \\ 2 \end{bmatrix}, \quad \mathrm{X}_{privot} = \begin{bmatrix} x_1 \\ x_2 \end{bmatrix}, \quad \mathrm{X}_{free} = x_3, \quad \mathrm{b}_1 \begin{bmatrix} 5 \\ 7 \end{bmatrix}$$

Here X_{privot} is a solution corresponding to the pivot columns and X_{free} represents free variables corresponding to the non-pivot columns, which are free variable that can take any value.

Solving for X_{privot}, we get:

$$\mathrm{Ix}_{pivot} + \mathrm{Fx}_{free} = \mathrm{b}_1, \qquad \mathrm{x}_{privot} = \mathrm{b}_1 - \mathrm{Fx}_{free} = \begin{bmatrix} 5 \\ 7 \end{bmatrix} + x_3 \begin{bmatrix} -1 \\ -2 \end{bmatrix}$$

If we let $\mathrm{x}_{free} = x_3 = 0$, we get a particular solution:

$$\mathrm{x}_{particular} = \begin{bmatrix} 5 \\ 7 \\ 0 \end{bmatrix}$$

But as the free variable x_3 can take any value c, the complete solution is:

$$\mathrm{x}_{comlete} = \begin{bmatrix} 5 \\ 7 \\ 0 \end{bmatrix} + x_3 \begin{bmatrix} -1 \\ -2 \\ 1 \end{bmatrix} = \mathrm{x}_{particular} + c\,\mathrm{x}_n = \mathrm{x}_{particular} + N(\mathrm{A})$$

where $\mathrm{x}_n = [-1, -2, 1]^T$ is the homogeneous solution for $\mathrm{Ax} = 0$, which is in the 1-D nullspace $N(\mathrm{A}) = c\,\mathrm{x}_n$. We therefore see that the complete solution of $\mathrm{Ax} = \mathrm{b}$ is the entire nullspace N(A) translated by the particular solution $\mathrm{x}_{particular}$. For example, when $c = 0,1,2$ we get the following solutions:

$$\mathrm{x} = \begin{bmatrix} 5 \\ 7 \\ 0 \end{bmatrix}, \quad \mathrm{x} = \begin{bmatrix} 4 \\ 5 \\ 1 \end{bmatrix}, \quad \mathrm{x} = \begin{bmatrix} 3 \\ 3 \\ 0 \end{bmatrix}$$

The following figure illustrates a 3-D R^3 space with a 2-D nullspace. The complete solution is the entire nullspace translated by a particular solution $\mathrm{x}_{particular}$.

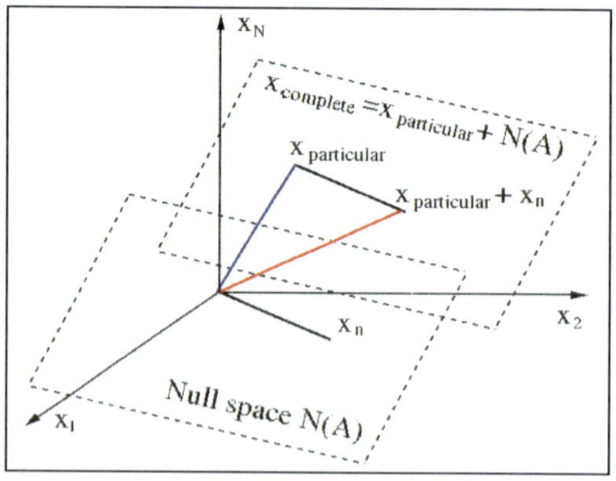

Example: Consider the following linear equation system with an $M \times N$ coefficient matrix A,

$$\mathrm{Ax} = \begin{bmatrix} 1 & 2 & 3 & 4 \\ 4 & 3 & 2 & 1 \\ -2 & 1 & 4 & 7 \end{bmatrix} \begin{bmatrix} x_1 \\ x_2 \\ x_3 \\ x_4 \end{bmatrix} = \begin{bmatrix} 3 \\ 2 \\ 4 \end{bmatrix} = \mathrm{b}$$

We find the complete solution of this system in the following steps:

1. Construct the augmented matrix and then convert it to the rref form:

$$\left[\begin{array}{cccc|c} 1 & 2 & 3 & 4 & 3 \\ 4 & 3 & 2 & 1 & 2 \\ -2 & 1 & 4 & 7 & 4 \end{array}\right] \to \left[\begin{array}{cccc|c} 1 & 2 & 3 & 4 & 3 \\ 0 & -5 & -10 & -15 & -10 \\ 0 & 5 & 10 & 15 & 10 \end{array}\right] \to \left[\begin{array}{cccc|c} 1 & 2 & 3 & 4 & 3 \\ 0 & 1 & 2 & 3 & 2 \\ 0 & 0 & 0 & 0 & 0 \end{array}\right]$$

$$\to \left[\begin{array}{cccc|c} 1 & 0 & -1 & -2 & -1 \\ 0 & 1 & 2 & 3 & 2 \\ 0 & 0 & 0 & 0 & 0 \end{array}\right] = \left[\begin{array}{cc|c} \mathrm{I} & \mathrm{F} & \mathrm{b}_1 \\ 0 & 0 & 0 \end{array}\right]$$

As there are two pivot rows in the rref, we know $R = rank(\mathrm{A}) = 2$. The equation system now takes this block matrix form:

$$\begin{bmatrix} \mathrm{I} & \mathrm{F} \\ 0 & 0 \end{bmatrix} \begin{bmatrix} \mathrm{x}_p \\ \mathrm{x}_f \end{bmatrix} = \begin{bmatrix} \mathrm{b}_1 \\ 0 \end{bmatrix}$$

Where,

$$I = \begin{bmatrix} 1 & 0 \\ 0 & 1 \end{bmatrix}, \ F = \begin{bmatrix} -1 & -2 \\ 2 & 3 \end{bmatrix}, \ x_p \begin{bmatrix} x_1 \\ x_2 \end{bmatrix}, \ x_f = \begin{bmatrix} x_3 \\ x_4 \end{bmatrix}, \ b_1 = \begin{bmatrix} -1 \\ 2 \end{bmatrix}$$

Multiplying out we get:

$$x_p + Fx_f = b_1$$

2. Find the homogeneous solution for equation $Ax = 0$ by setting $b_1 = 0$:

$$Ix_p + Fx_f = 0, \text{ i.e. } \quad x_p = -Fx_f, \quad \text{i.e. } \begin{bmatrix} x_1 \\ x_2 \end{bmatrix} = \begin{bmatrix} 1 & 2 \\ -2 & -3 \end{bmatrix} \begin{bmatrix} x_3 \\ x_4 \end{bmatrix}$$

We use the two standard basis vectors $x_f = [x_3, x_4]^T = [1\,0]^T$ and $x_f = [x_3, x_4]^T = [0\,1]^T$ of the nullspace $N(A)$ and get:

$$x_p = \begin{bmatrix} x_1 \\ x_2 \end{bmatrix} = \begin{bmatrix} 1 \\ -2 \end{bmatrix} \text{ or } \begin{bmatrix} 2 \\ -3 \end{bmatrix} \quad \text{i.e. } x_{n1} = \begin{bmatrix} x_1 \\ x_2 \\ x_3 \\ x_4 \end{bmatrix} = \begin{bmatrix} 1 \\ -2 \\ 1 \\ 0 \end{bmatrix}, \quad x_{n2} = \begin{bmatrix} x_1 \\ x_2 \\ x_3 \\ x_4 \end{bmatrix} = \begin{bmatrix} 2 \\ -3 \\ 0 \\ 1 \end{bmatrix}$$

3. Find the particular solution of the non-homogeneous equation $Ax = b$ by setting $x_f = 0$

$$Ix_p + Fx_f = x_p = b_1, \text{ i.e. } \begin{bmatrix} x_1 \\ x_2 \end{bmatrix} = \begin{bmatrix} -1 \\ 2 \end{bmatrix} \quad \text{i.e. } x_{particular} \begin{bmatrix} -1 \\ 2 \\ 0 \\ 0 \end{bmatrix}$$

4. Get the complete solution:

$$x_{comlete} = x_{particular} + N(A) = x_{particular} + c_1 x_{n1} + c_2 + x_{n2} = \begin{bmatrix} -1 \\ 2 \\ 0 \\ 0 \end{bmatrix} + c_1 \begin{bmatrix} 1 \\ -2 \\ 1 \\ 0 \end{bmatrix} + c_2 \begin{bmatrix} 2 \\ -3 \\ 0 \\ 1 \end{bmatrix}$$

If the right-hand side is $b' = [1, 3, 5]^T$, then the row reduction of the augmented matrix yields:

$$\left[\begin{array}{cccc|c} 1 & 2 & 3 & 4 & 1 \\ 4 & 3 & 2 & 1 & 3 \\ -2 & 1 & 4 & 7 & 5 \end{array}\right] \to \left[\begin{array}{cccc|c} 1 & 2 & 3 & 4 & 3 \\ 0 & -5 & -10 & -15 & -1 \\ 0 & 5 & 10 & 15 & 7 \end{array}\right] \to \left[\begin{array}{cccc|c} 1 & 2 & 3 & 4 & 1 \\ 0 & 1 & 2 & 3 & 0.2 \\ 0 & 0 & 0 & 0 & 1.2 \end{array}\right]$$

The equation corresponding to the last row is $0 = 1.2$, indicating the system is not solvable (even though the coefficient matrix does not have full rank), as $b' = [1, 3, 5]^T$ on the right-hand is not in the image, the column space $C(A)$.

Example: Consider the linear equation system with coefficient matrix A^T, the transpose of the coefficient matrix in the previous example,

$$A^T y = \begin{bmatrix} 1 & 4 & -2 \\ 2 & 3 & 1 \\ 4 & 1 & 7 \end{bmatrix} \begin{bmatrix} y_1 \\ y_2 \\ y_3 \end{bmatrix} = \begin{bmatrix} 3 \\ 11 \\ 19 \\ 27 \end{bmatrix} = c$$

1. Carrying out elementary row operations to the augmented matrix we get:

$$\left[\begin{array}{ccc|c} 1 & 4 & -2 & 3 \\ 2 & 3 & 1 & 11 \\ 3 & 2 & 4 & 19 \\ 4 & 1 & 7 & 27 \end{array}\right] \rightarrow \left[\begin{array}{ccc|c} 1 & 4 & -2 & 3 \\ 0 & -5 & 5 & 5 \\ 0 & -10 & 10 & 10 \\ 0 & -15 & 15 & 15 \end{array}\right] \rightarrow \left[\begin{array}{ccc|c} 1 & 4 & -2 & 3 \\ 0 & 1 & -1 & -1 \\ 0 & 0 & 0 & 0 \\ 0 & 0 & 0 & 0 \end{array}\right] \rightarrow \left[\begin{array}{ccc|c} 1 & 0 & 2 & 7 \\ 0 & 1 & -1 & -1 \\ 0 & 0 & 0 & 0 \\ 0 & 0 & 0 & 0 \end{array}\right]$$

The original system can now be written as:

$$\begin{bmatrix} I & F \\ 0 & 0 \end{bmatrix} \begin{bmatrix} x_p \\ x_f \end{bmatrix} = \begin{bmatrix} c_1 \\ 0 \end{bmatrix}$$

where,

$$I = \begin{bmatrix} 1 & 0 \\ 0 & 1 \end{bmatrix}, \quad F = \begin{bmatrix} 2 \\ -1 \end{bmatrix}, \quad y_p = \begin{bmatrix} y_1 \\ y_2 \end{bmatrix}, \quad y_f = y_3, \quad c_1 = \begin{bmatrix} 7 \\ -1 \end{bmatrix}$$

Multiplying out we get:

$$Iy_p + Fy_f = c_1$$

2. Find the homogeneous solution for $A^T y = 0$, by setting $c_1 = 0$. Assuming the free variable $y_f = y_3 = 1$, We get:

$$Iy_p + Fy_f = 0, \text{ i.e. } y_p = -Fy_f = \begin{bmatrix} y_1 \\ y_2 \end{bmatrix} = -\begin{bmatrix} 2 \\ -1 \end{bmatrix} y_3 = \begin{bmatrix} -2 \\ 1 \end{bmatrix}, \text{ i.e. } y_n = \begin{bmatrix} -2 \\ 1 \\ 0 \end{bmatrix}$$

3. Find the particular solution of $A^T y = c$ by assuming $y_f = y_3 = 0$:

$$Iy_p + Fy_f = y_p = c_1 = \begin{bmatrix} 7 \\ -1 \end{bmatrix}, \quad \text{i.e. } y_{particular} = \begin{bmatrix} 7 \\ -1 \\ 0 \end{bmatrix}$$

4. Get the complete solution:

$$\mathbf{y}_{complete} = \mathbf{y}_{particular} + N(\mathrm{A}^T) = \mathbf{y}_{particular} + c\mathbf{y}_3 = \begin{bmatrix} 7 \\ -1 \\ 0 \end{bmatrix} + c\begin{bmatrix} -2 \\ 1 \\ 1 \end{bmatrix}$$

If the right-hand side is $\mathrm{c}' = [1,1,2,3]^T$,

$$\left[\begin{array}{ccc|c} 1 & 4 & -2 & 1 \\ 2 & 3 & 1 & 1 \\ 3 & 2 & 4 & 2 \\ 4 & 1 & 7 & 3 \end{array}\right] \rightarrow \left[\begin{array}{ccc|c} 1 & 4 & -2 & 1 \\ 0 & -5 & 5 & -1 \\ 0 & -10 & 10 & -1 \\ 0 & -15 & 15 & -1 \end{array}\right] \rightarrow \left[\begin{array}{ccc|c} 1 & 4 & -2 & 1 \\ 0 & 1 & -1 & 1/5 \\ 0 & 0 & 0 & 1 \\ 0 & 0 & 0 & 2 \end{array}\right]$$

indicating the system is not solvable, as this $\mathrm{c}' \notin C(\mathrm{A}^T)$.

All four subspaces associated with this matrix $\underline{\mathrm{A}}$ with $M = 3, N = 4,$ and $R = 2$ can be obtained in the last two examples:

1. The row space $R(\mathrm{A}) = C(\mathrm{A}^T)$ is an R-D subspace of R^4, spanned by the firs $R = 2$ pivot rows of the rref of A :

$$R(\mathrm{A}) = C(\mathrm{A}^T) = span\left(\begin{bmatrix} 1 \\ 0 \\ -1 \\ -2 \end{bmatrix}, \begin{bmatrix} 0 \\ 1 \\ 2 \\ 3 \end{bmatrix}\right)$$

2. The nullspace $N(\mathrm{A})$ is an (N-R)-D subspace of R^4 spanned by the $N - R$ independent homogeneous solutions:

$$N(\mathrm{A}) = span\left(\begin{bmatrix} 1 \\ -2 \\ 1 \\ 0 \end{bmatrix}, \begin{bmatrix} 2 \\ -3 \\ 0 \\ 1 \end{bmatrix}\right)$$

Note that the basis vectors of $N(\mathrm{A})$ are indeed orthogonal to those of $R(\mathrm{A})$.

3. The column space of A is the same as the row space of A^T, which is the R-D subspace of R^3 spanned by the first two pivot rows of the rref of A^T.

$$C(\mathrm{A}) = R(\mathrm{A}^T) = span\left(\begin{bmatrix} 1 \\ 0 \\ 2 \end{bmatrix}, \begin{bmatrix} 0 \\ 1 \\ -1 \end{bmatrix}\right)$$

4. The left nullspace $N(A^T)$ is a (M-R)-D subspace of R^3 spanned by the homogeneous solutions (here one $3-2=1$ solution):

$$N(A^T) = span\left(\begin{bmatrix} -2 \\ 1 \\ 1 \end{bmatrix}\right) = c\begin{bmatrix} -2 \\ 1 \\ 1 \end{bmatrix}$$

Again note that the basis vectors of $N(A^T)$ are orthogonal to those of $C(A)$.

In general, here are the ways to find the bases of the four subspaces:

1. The basis vectors of $R(A)$ are the pivot rows of the rref form of A.
2. The basis vectors of $C(A)$ are the pivot rows of the rref form of A^T.
3. To find the basis for $N(A)$, reduce A to the rref form, identify all free variables x_f corresponding to non-pivot columns, set one of them to 1 and all others to 0, solve homogeneous system $Ax = 0$ to find the pivot variables x_p o get one basis vector. Repeat the process for each of the free variables to get all basis vectors.
4. By doing the same as above for A^T, we get the basis for $R(A^T)$.

Note that while the basis of $R(A)$ can be obtained from the rref form of Aas its rows are equivalent to those of A, the basis of $C(A)$ cannot be obtained as the columns of the rref form, as the columns of A have been changed by the row deduction operations and are not equivalent to the columns of the resulting rref form. But the basis of $C(A)$ can be obtained from the rref form of A^T, as its rows are equivalent to those of A^T, which are the columns of A.

We make the following observations:

1. The basis vectors of each of the four spaces are independent, the basis vectors of $R(A)$ and $N(A)$ are orthogonal, and $dim\, R(A) + dim\, N(A) = M$. Similarly, the basis vectors of $C(A)$ and A^T are orthogonal, and $dim\, C(A) + dim\, N(A^T) = N$. In other words, the four subspaces indeed satisfy the following orthogonal and complementary properties:

$$C(A) \perp N(A^T), \quad R(A) \perp N(A), \quad C(A) \oplus N(A^T) = R^3, \quad R(A) \oplus N(A) = R^4$$

i.e., they are orthogonal complements: $C(A) = N(A)^{\perp}, \quad R(A) = N(A)^{\perp}$.

2. For $Ax = b$ to be solvable, the constant vector on the right-hand side must be in the column space, i.e., $b \in C(A)$. Otherwise the equation is not solvable, even when $R < M < N$. Similarly, for $A^T x = c$ to be solvable, c must be in the row space $c \in R(A)$. In the examples above, both b and c are indeed in their corresponding column spaces:

$$b = \begin{bmatrix} 3 \\ 2 \\ 4 \end{bmatrix} = 3\begin{bmatrix} 1 \\ 0 \\ 2 \end{bmatrix} + 2\begin{bmatrix} 0 \\ 1 \\ -1 \end{bmatrix} \in C(A), \quad c = \begin{bmatrix} 3 \\ 11 \\ 19 \\ 27 \end{bmatrix} = 3\begin{bmatrix} 1 \\ 0 \\ -1 \\ -2 \end{bmatrix} + 11\begin{bmatrix} 0 \\ 1 \\ 2 \\ 3 \end{bmatrix} \in R(A)$$

But as $b' = [1,3,5]^T \notin C(A)$ and $c' = [1,1,2,3]^T \notin R(A)$, the corresponding systems have no solutions.

3. All homogeneous solutions of $Ax = 0$ are in the nullspace $x_p \in R(A)$, but in general the particular solutions $x_p \in R(A) \oplus N(A) = R^N$ are not necessarily in the row space $x_p \in R(A)$. In the example above, is a linear combination of the $R = 2$ basis vectors of (A) and $N - R = 2$ basis vectors of $N(A)$:

$$x_p = \begin{bmatrix} -1 \\ 2 \\ 0 \\ 0 \end{bmatrix} = \left(0.1 \begin{bmatrix} 1 \\ 0 \\ -1 \\ -2 \end{bmatrix} + 0.2 \begin{bmatrix} 0 \\ 1 \\ 2 \\ 3 \end{bmatrix} \right) + \left(-0.3 \begin{bmatrix} 1 \\ -2 \\ 1 \\ 0 \end{bmatrix} - 0.4 \begin{bmatrix} 2 \\ -3 \\ 0 \\ 1 \end{bmatrix} \right) = x'_p + x''_p$$

where,

$$x'_p = \begin{bmatrix} 0.1 \\ 0.2 \\ 0.3 \\ 0.4 \end{bmatrix} \in R(A), \quad x''_p = \begin{bmatrix} -1.1 \\ 1.8 \\ -0.3 \\ -0.4 \end{bmatrix} \in N(A)$$

are the projections of x_p onto $R(A)$ and $N(A)$, respectively, and x'_p is another particular solution without any homogeneous component that satisfies $Ax = b$, while x''_p is a homogeneous solution satisfying $Ax = 0$.

4. All homogeneous solutions of $A^T y = c$ are in the left nullspace $y_n \in N(A^T)$, but in general the particular solutions $y_p \in C(A) \oplus N(A^T) = R^M$ are not necessarily in the column space. In the example above, $y_p \in R^M$ is a linear combination of the $R = 2$ basis vectors of $C(A)$ and $M - R = 1$ basis vector of $N(A^T)$:

$$y_p = \begin{bmatrix} 7 \\ -1 \\ 0 \end{bmatrix} = \left(2 \begin{bmatrix} 1 \\ 0 \\ 2 \end{bmatrix} + 1.5 \begin{bmatrix} 0 \\ 1 \\ -1 \end{bmatrix} \right) - 2.5 \begin{bmatrix} -2 \\ 1 \\ 1 \end{bmatrix} = y'_p + y''_p$$

where, $y'_p = [2, 1.5, 2.5]^T \in C(A)$ is a particular solution (without any homogeneous component) that satisfies $A^T y = c$.

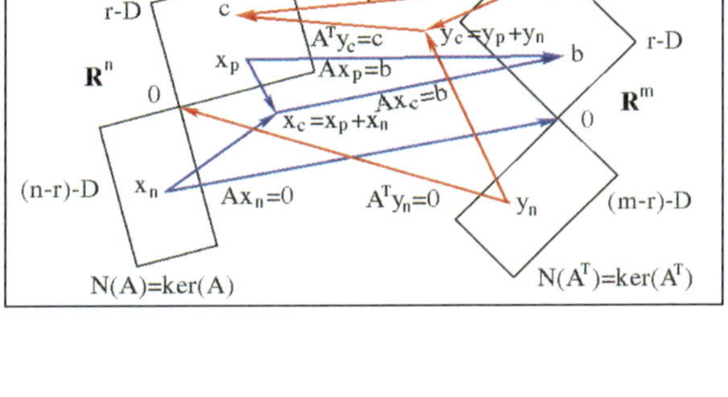

All these results can be illustrated in the figure below, where we assume $x_p \in R(A)$ and $y_p \in C(A)$, which is not the case in these examples, but will be true when the systems were solved by the SVD method.

Here is a summary of the four subspaces associated with an M by N matrix $\underline{A}$ of rank R.

M, N, R	dim R(A)	dim N(A)	dim C(A)	dim R(A^T)	solvable, Ax=b
M=N=R	R	0	R	0	solvable, $x=A^{-1}b$ is unique solution
M>N=R	R	0	R	M-R	over-constrained, solvable if $b \in C(A)$
N>M=R	R	N-R	R	0	under-constrained, solvable, infinite solutions $x = x_p + N(A)$
R<min(M, N)	R	N-R	R	M-R	solvable only if $b \in C(A)$, infinite solutions

The following figure illustrates the case when $M = N = 3$, and when $b \notin C(A)$. he system can only be approximately solved for the projection $p \in$ (A) of b in the column space. This is the optimal approximation in the least-squares sense as the error is minimized $||e|| \to min$.

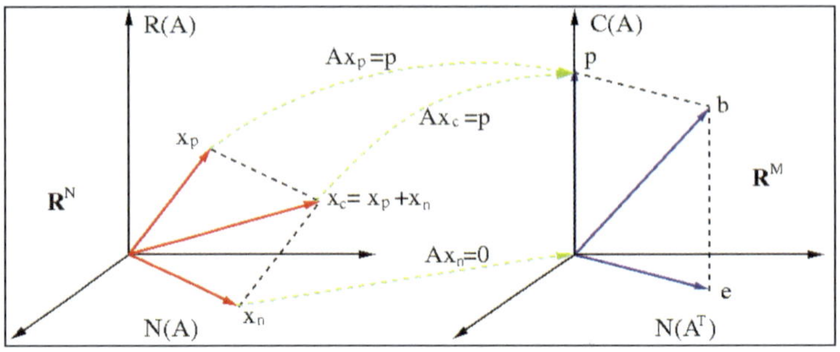

Cayley-Hamilton Theorem

Let $C^{n\times m}$ be the set of complex $(n\times m)$ matrices and $A \in C^{n\times n}$. Let,

$$\varphi(s) = \det[I_n\lambda - A] = \sum_{i-0}^{n} a_i\lambda^i (a_n = 1)$$

be the characteristic polynomial of A, where I_n is the $(n\times n)$ identity matrix. The Cayley–Hamilton theorem says [a2], [a9] that every square matrix satisfies its own characteristic equation, i.e.

$$\varphi(A) = \sum_{i=0}^{m} a_i A^i = 0,$$

where 0 is the zero-matrix.

The classical Cayley–Hamilton theorem can be extended to rectangle matrices. A matrix $A \in C^{m\times n}$ for $n > m$ may be written as $A = [A_1, A_2]$, $A_1 \in C^{m\times m}$, $A_2 \in C^{m\times(n-m)}$. Let,

$$\det[I_n\lambda - A_1] = \sum_{i=0}^{m} a_i\lambda^i (a_m = 1).$$

Then, the matrix $A \in C^{m\times n}(n > m)$ satisfies the equation:

$$\sum_{i=0}^{m} a_m - i\left[A_1^{m-i}, A_1^{n-i-1}A_2\right] = 0.$$

A matrix $A \in C^{m\times n}(m > n)$ may be written as:

$$A = \begin{bmatrix} A_1 \\ A_2 \end{bmatrix}, \; A_1 \in C^{n\times n}, \; A_2 \in C^{(m-n)\times n}.$$

Let,

$$\det[I_n\lambda - A_1] = \sum_{i=0}^{n} a_i\lambda^i (a_n = 1).$$

Then the matrix $A \in C^{m\times n}(m > n)$ satisfies the equation:

$$\sum_{i=0}^{n} a_{n-1}\begin{bmatrix} A^{m-i} \\ A_2A_1^{m-i-1} \end{bmatrix} = 0_{mn}.$$

The Cayley–Hamilton theorem can be also extended to block matrices. Let,

$$A_i = \begin{bmatrix} A_{11} & \cdots & A_{1m} \\ \cdots & \cdots & \cdots \\ A_{m1} & \cdots & A_{mm} \end{bmatrix} \in C^{mn\times mn},$$

where, $A_{ij} \in C^{n\times n}$ are commutative, i.e. $A_{ij} = A_{kl} = A_{kl}A_{ij}$ for all $i, j, k = 1, \cdots, m$ Let,

$$\Delta(\wedge) = \mathrm{Det}[l_m \otimes \wedge - A_1] =$$
$$= \wedge^m + D_1\wedge^{m-1} + \cdots + D_{m-1}\wedge + D_m, \; D_k \in C^{n\times n}, \; k = 1, \cdots, m,$$

be the matrix characteristic polynomial and let $\wedge \in C^{n\times n}$ be the matrix (block) eigenvalue of A_1 where $\otimes$ denotes the Kronecker product. The matrix $\Delta(\wedge)$ is obtained by developing the determinant of $[l_m \otimes \wedge - A_1]$ considering its commuting blocks as elements.

The block matrix satisfies the equation:

$$\Delta(A_1) = \sum_{i=0}^{m} (l_m \otimes D_{m-i})A_1^i = 0(D_0 = I_n).$$

Consider now a rectangular block matrix $A = [A_l, A_2] \in C^{mn\times(mn+p),}$ where A_1 has the form and $A_2 \in C^{mn\times p(p>0)}$. The matrix A satisfies the equation:

$$\sum_{i=0}^{m} (I_m \otimes D_{m-i})[A_1^{i+1}, A_1^i A_2] = 0(D_0 = I_n).$$

If,

$$A=\begin{bmatrix} A_1 \\ A_2 \end{bmatrix} \in C^{(mn+p)\times m},$$

where A_1 has the form (a1) and $A_2 \in C^{p\times mn}$, then:

$$\sum_{i=0}^{m} \begin{bmatrix} A_1 \\ A_2 \end{bmatrix} (I_m \otimes D_{m-i}) A_1^i = 0(D_0 - l_n).$$

A pair of matrices $E, A \in C^{n\times n}$ is called regular if $\det[E\lambda - A] \neq 0$ for some $\lambda \in \mathrm{C}$. The pair is called standard if there exist scalars $\alpha, \beta \in \mathrm{C}$ such that $E\alpha + A\beta = I_n$ If the pair $E, A \in C^{n\times n}$ is regular, then the pair:

$$E = [E\lambda - A]^{-1} E, \bar{A} = [E\lambda - A]^{-1} A$$

is standard. If the pair $E, A \in C^{n\times n}$ is standard, then it is also commutative ($EA = AE$). Let a pair $E, A \in C^{n\times n}$ be standard (commutative) and,

$$\Delta(\lambda, \mu) = \det[E\lambda - A\mu] = \sum_{i=0}^{n} a_{i, n-1} \lambda^i \mu^{n-i}.$$

Then the pair satisfies the equation:

$$\Delta(A, E) = \sum_{i=0}^{n} a_{i, n-i} A^i E^{n-i} = 0.$$

In a particular case, with $\det[E\lambda - A] = \sum_{i=0\ a_i s^i}^{n}$, if follows that

$$\sum_{i=0}^{n} a_i A^i E^{n-i} = 0.$$

Let $P_n(C)$ be the set of n - order square complex matrices that commute in pairs and let $M_m(P_n)$ be the set of square matrices partitioned in m^2 blocks belonging to $P_n(C)$.

Consider a standard pair of block matrices $E, A \in M_m(P_n)$ and let the matrix polynomial:

$$\Delta(\Delta, M) = \mathrm{Det}[E \otimes \wedge - A \otimes M] =$$

$$= \sum_{i=0}^{m} D_{i, m-1} \wedge^i M^{m-i}, D_{ij} \in C^{n\times n},$$

be its matrix characteristic polynomial. The pair $(\wedge, M)$ is called the block-eigenvalue pair of the pair E, A. Then:

$$\Delta(A, E) = \sum_{i=0}^{m} I \otimes D_{i,n-i} A^i E^{m-i} = 0.$$

The Cayley–Hamilton theorem can be also extended to singular two-dimensional linear systems described by Roesser-type or Fomasini–Marchesini-type models. The singular two-dimensional Roesser model is given by:

$$\begin{bmatrix} E_1 & E_2 \\ E_3 & E_4 \end{bmatrix}\begin{bmatrix} x^h_{i+1,j} \\ x^v_{i,j+1} \end{bmatrix} = \begin{bmatrix} A_1 & A_2 \\ A_3 & A_4 \end{bmatrix}\begin{bmatrix} x^h_{ij} \\ x^v_{ij} \end{bmatrix} + \begin{bmatrix} B_1 \\ B_2 \end{bmatrix}\mu_{ij},$$

$$i, j, \in Z+.$$

Here, Z_+ is the set of non-negative integers; $x^h_{ij} \in R^{n_1}$, respectively $x^v_{ij} \in R^{n_2}$, are the horizontal, respectively vertical, semi-state vector at the point $(i, j)\mu_{ij} \in R^m$ is the input vector;

$$E_k, A_k (k = 1, \cdots, 4)$$

and $B_i (i = 1, 2)$ have dimensions compatible with x^h_{ij} and x^v_{ij}:

$$\begin{bmatrix} E_1 & E_2 \\ E_3 & E_4 \end{bmatrix}$$

may be singular. The characteristic polynomial has the form:

$$\Delta(z_1, z_2) = \det\begin{bmatrix} E_{1Z1} - A_1 & E_{2Z2} - A_2 \\ E_{3Z1} - A_3 & E_{4Z2} - A_4 \end{bmatrix} = \sum_{i=0}^{r_1}\sum_{j=0}^{r_2} a_{ij} z_1^i z_2^j$$

and the transition matrices $T_{p}, _{q,p,q} \in z_+$, are defined by:

$$\begin{bmatrix} E_l & 0 \\ E_3 & 0 \end{bmatrix} T_{p,q-1} + \begin{bmatrix} 0 & E_2 \\ 0 & E_4 \end{bmatrix} T_{p-1,q} +$$

$$+ \begin{bmatrix} A_1 & A_2 \\ A_3 & A_4 4 \end{bmatrix} T_{p-l,q-1} =$$

$$= \begin{cases} I_n, & p = q = 0, \\ 0, & p \neq 0 \text{ or/ and } q \neq 0. \end{cases}$$

If $E = I_n, n_l + n_2$ (the standard Roesser model), then the transition matrices T_{pq} may be computed recursively, using the formula $T_{pq} = T_{10} T_{p-1,q} + T_{01} T_{p,q-1}$ where $T_{00} = I_n$,

$$T_{10} = \begin{bmatrix} A_1 & A_2 \\ 0 & 0 \end{bmatrix}, T_{01} = \begin{bmatrix} 0 & 0 \\ A_3 & A_4 \end{bmatrix}.$$

The matrices T_{pq} satisfy the equation:

$$\sum_{i=0}^{r_1}\sum_{j=0}^{r_2} a_{ij} T_{ij} = 0$$

The singular two-dimensional Fornasini–Marchesini model is given by:

$$E_{xi+1,j+1} = A_0 x_{ij+1,j} + A_1 x_{i+1,j} + A_2 x_{i,j+1} + B_{uij},$$
$$i, j \in Z_+,$$

where, $x_{ij} \in R^n$ is the local semi-vector at the point $(i,j)\mu_j^i \in R^m$ is the input vector, $E, A_k \in R^{n\times m}$ and E is possibly singular. The characteristic polynomial has the form:

$$\Delta(z_l, z_2) = \det[E_{z_1 z_2} - A_1 z_1 - A_2 z_2 - A_0] = \sum_{i=0}^{r_1}\sum_{i=0}^{r_2} a_{ij} z_1^i z_2^j$$

and the transition matrices $T_{p,q,p,q} \in Z_+$, are defined by:

$$ET_{pq} - A_0 T_{p-1,q-1} - A_1 T_{p,q-1} - A_2 T_{p-1,q} =$$
$$= \begin{cases} I_n, & p = q = 0, \\ 0, & p \neq 0 \text{ or/ and } q \neq 0. \end{cases}$$

The matrices T_{pq} satisfy the equation:

$$\sum_{i=0}^{r_1}\sum_{i=0}^{r_2} a_{ij} T_{ij} = 0$$

The theorems may be also extended to two-dimensional continuous-discrete linear systems.

Chebotarëv's Theorem on Roots of Unity

Let p be a prime and ω a primitive pth root of unity. Then all minors of the matrix $\left(\omega^{ij}\right)_{i,j=0}^{p-1}$ are non-zero.

Let p be a prime and ω a primitive pth root of unity. We write F_p for the field with p elements.

Theorem: For any sets $I, J \subseteq F_p$ with equal cardinality, the matrix $\left(\omega^{ij}\right)_{i\in I, j\in J}$ has non-zero determinant.

Independent proofs were given by Dieudonné, Evans and Isaacs, and Terence Tao. Tao points out that the theorem is equivalent to the inequality $|\text{supp} f| + |\text{supp} \hat{f}| \geq p+1$ holding for any function $\not\equiv f : F \to C$ and its Fourier transform $\hat{f}$, a fact also discovered independently by András Biró.

We give a very simple proof via the following two lemmas. Lemma is covered by, but we include a proof for the sake of completeness.

Lemma: $Z[\omega]/(1-\omega) = F_p$.

Proof: Let Ω be an indeterminate and let,

$$\Phi_p(\Omega) = 1 + \Omega + \cdots + \Omega^{p-1}$$

be the minimal polynomial of the algebraic integer ω. Consider the surjective ring homomorphisms:

$$Z[\Omega] \to Z[\Omega]/(\Phi_p(\Omega)) = Z[\omega], \quad \Omega \mapsto \omega$$

and

$$Z[\Omega] \to Z[\Omega]/(1-\Omega, p) = F_p, \quad \Omega \mapsto 1.$$

The latter kernel contains the former one since $\Phi_p(\Omega) \equiv p$ mod (1-Ω). Therefore, the latter homomorphism factors through the former one via a surjective homomorphism $Z[\omega] \to F_p$ whose kernel is the ideal:

$$(1-\Omega, p)/(\Phi_p(\Omega)) = (1-\omega, p) = (1-\omega),$$

the last equality following from $p \equiv \Phi_p(\omega) = 0 \bmod (1-\omega)$.

Lemma: Let $0 \not\equiv g(x) \in F_p[x]$ be a polynomial of degree $< p$. Then the multiplicity of any element $0 \neq a \in F_p$ as a root of $g(x)$ is strictly less than the number of non-zero coefficients of $g(x)$.

Proof: For $g(x)$ constant, the lemma is obviously true. Assume that it is true for any $g(x)$ of degree $< k$, with some fixed $1 \le k < p$, and take $g(x)$ of degree k If $g(0) = 0$, then $g(x)$ has the same number of non-zero coefficients and the same multiplicity of vanishing at a as $g(x)/x$ does, so the lemma is true for $g(x)$ If $g(0) \neq 0$, then the number of non-zero coefficients exceeds the corresponding number for the derivative $g'(x)$ by 1, and the multiplicity of vanishing at a exceeds that of $g'(x)$ by at most 1. Now $g'(x) \not\equiv 0$ since $g(x)$ is of positive degree $k < p$, , so the inequality of the lemma holds for $g'(x)$ and therefore also for $g(x)$.

Proof of the theorem: The theorem is equivalent to saying that if numbers $a_j \in Q(\omega)$ $(j \in J)$ satisfy $\sum_{j \in J} a_j \omega^{ij} = 0$ for all $i \in I$, then all a_j must be zero. In fact, we may clearly assume that aj ∈ Z[ω]. The above equalities mean that the polynomial:

$$g(x) = \sum_{j \in J} a_j x^j \in Z[\omega, x]$$

vanishes at ω^i for all $i \in I$. So $g(x)$ is divisible by $\prod_{i \in I}(x - \omega^i)$. Applying the homomorphism

$$Z[\omega] \to Z[\omega]/(1-\omega) = F_p$$

to the coefficients of $g(x)$ we get a polynomial $\bar{g}(x) \in F_p[x]$ that is divisible by $(x-1)^{|I|}$. On the other hand, $\bar{g}(x)$ has at most $|J|$ non-zero coefficients. As $|I| = |J|$, we deduce from Lemma 2 that $\bar{g}(x) \equiv 0$. This means that all a_j are divisible by 1 − ω. We may divide all of them by 1 − ω and iterate the argument. This leads to descent infinite unless all a_j are zero.

The Remainder Theorem and the Factor Theorem

Remainder Theorem

If $p(x)$ is any polynomial, then the remainder after division by $(x-a)$ is exactly $p(a)$.

Proof: Dividing $p(x)$ by $(x-a)$, we get a quotient $q(x)$ and a remainder R, which must be a constant. We can then write $p(x)$ as:

$$p(x)=(x-a)\cdot q(x)+R$$

We now simply plug in x = a in the above equation.

$$p(a) = (a - a) \cdot q(x) + R$$
$$p(a) = R$$

and have the desired result.

Factor Theorem

The value a is a root of the polynomial $p(x)$ if and only if $(x-a)$ is a factor of $p(x)$.

Proof:

1. ($\Rightarrow$) Assume that a a is a root of the polynomial $p(x)$. This means that $p(a) = 0$.

   By the remainder theorem, we conclude that the remainder after division by $(x-a)$ must be zero. So,

   $$p(x) = (x - a) \cdot q(x) + 0$$
   $$p(x) = (x - a) \cdot q(x)$$

   Therefore, $(x-a)$ is a factor of $p(x)$.

2. ($\Leftarrow$) Now we assume that $(x-a)$ is a factor of $p(x)$. Then,

   $$p(x) = (x - a) \cdot q(x)$$
   $$p(a) = (a - a) \cdot q(x) = 0$$

   So, a is a root of $p(x)$.

Example: Evaluate,

$$\int \frac{x^2 + x}{x^3 - 1} dx$$

In order to apply partial fractions, we need to factor the denominator,

$$p(x) = x^3 - 1$$

completely.

We of course immediately notice that 1 is a root of this polynomial. By the Factor Theorem, we can then conclude that $(x - 1)$ must be a factor. Dividing, we get:

$$p(x) = (x - 1)(x^2 + x + 1)$$

Now we need to try to factor the remaining quadratic term. As we did above, we try to find factors by finding roots. However, this quadratic does not have any roots, because it is always positive; we can see this most easily by completing the square:

$$x^2 + x + 1 = \left(x + \frac{1}{2}\right)^2 + \frac{3}{4} \geq \frac{3}{4}$$

Since it has no roots, the Factor Theorem then tells us that it can have no linear factors; and since it is quadratic, any factor must be linear. So, it can therefore have no factors at all. So it is irreducible.

So, to evaluate the anti-derivative, we would first apply the partial fractions technique to:

$$\frac{x^2 + x}{(x - 1)(x^2 + x + 1)}$$

Fundamental Theorem of Ring Homomorphisms

If $(R,+,*)$ is a ring and $I,+,*)$ is an ideal, then the quotient of R by I is the set R/I of equivalence classes $a+I=\{a+i : i \in I\}$ with the operations of addition and multiplication defined for all $(a+I),(b+I) \in R/I$ by:

$$(a+I)+(b+I)=(a+b)+I$$
$$(a+I)*(b+I)=(a*b)+I$$

So in order to define a quotient ring we need a ring $(R,+,*)$ and an ideal $(I,+,*)$. Now suppose that R and S are homomorphic rings with homomorphism $\phi : R \to S$. Then $\ker(\phi)$ is a subring of R (this is relatively easy to show). We would ultimately like $\ker(\phi)$ o be an ideal of R to define the quotient $R/\mathrm{k}(\phi)$.

This is not true in general though. Recall that when we defined a ring, we required that there exists an element $1 \in R$ such that $a*1=a$ and $1*a=a$ for all $a \in R$, that is, we required the existence of a multiplicative identity. However, by definition, $\ker(\phi)$ in general cannot be a subring of R because if 1 is the identity of R, then by definition of a ring homomorphism, $\phi(1)$ is mapped to the multiplicative identity of S and not to the additive identity of S.

To establish a fundamental theorem of ring homomorphisms, we make a small exception in not requiring that $\ker(\phi)$ is an ideal for the quotient $R / \mathrm{k}(\phi)$ to be defined.

Theorem:

The Fundamental Theorem of Ring Homomorphisms: Let,

$$(R,+_{1,*1})\text{and}(S,+_{2,*2})$$

be homomorphic rings with ring homomorphism $\phi: R \to S$. Then $R / \ker(\phi) \cong \phi(R)$.

- Proof: Let $K = \ker(\phi)$ and $\psi : R / K \to \phi(R)$ be defined for all $(a+K) \in R$ by :

  $\psi(a+K) = \phi(a)$

- Then ψ is well-defined, for if $a+K = b+K$ then $a = b+k$ for some $k \in K$ and so:

  $\psi(a+K) = \psi((b+k)+K) = \psi((b+K)+K) = \psi(b+K)$

- Now let $(a+K),(b+K) \in R / K$. then:

  $\psi((a+K)+(b+K)) = \psi((a+b)+K) = \phi(a+b) = \phi(a)+\phi(b) = \psi(a+K)+\psi(b+K)$
  $\psi((a+K)*(b+K)) = \psi((a*b)+K) = \phi(a*b) = \phi(a)*\phi(b) = \psi(a+K)*\psi(b+K)$

- We lastly show that ψ is bijective.

- Let $(a+K),(b+K) \in R / K$ and suppose that $\psi(a+K) = \psi(b+K)$. Then $\phi(a) = \phi(b)$. So $\phi(a-b) = 0$. So $a-b \in K$. So $a+K = b+K$. Hence ψ is injective.

- Furthermore, for all $a \in \phi(R)$ we have that $(a+I) \in R / K$ is such that $\psi(a+K) = a$. So ψ is surjective.

- Hence ψ is bijective and so ψ is an isomorphism from $\psi : R / K \,\text{to}\, \phi(R)$ that is:

  $R / \ker(\phi) \cong \phi(R)$

For example, consider the rings $\mathbb{Z}$ and $\mathbb{Z}_n$. We can define a homomorphism $\phi: \mathbb{Z} \to \mathbb{Z}_n$ for all $z \in \mathbb{Z}$ by:

$$\phi(z) = z \bmod n$$

Indeed ϕ is a homomorphism, as you should verify. Note that:

$$\ker\phi(z) = \{0, \pm n, \pm 2_n, ..\} = n\mathbb{Z}$$

So by the fundamental theorem of ring homomorphisms we have that:

$$\mathbb{Z} / n\mathbb{Z} \cong \mathbb{Z}_n$$

Fundamental Theorem of Galois Theory

Let L/K be a finite Galois extension. Then there is an inclusion reversing bijection between the subgroups of the Galois group Gal(L/K) and intermediary subfields L/M/K. Given a subgroup H, let M = L H and given an intermediary field L/M/K, let H = Gal(L/M).

Proof: This will be an easy consequence of all that has gone before. Suppose that we are given a subgroup H of G. Let M = L^H and then set K = Gal(L/M). We want to show that K = H. As we have already proved that:

$$H \subset K,$$

and |G| = [L : K] is finite, it suffices to prove that the cardinality of K and is at most the cardinality of H. But:

$$|H| = [L : M],$$

and there are at most [L : M] automorphisms of L/M, so that:

$$|K| \le [L : M] = |H|.$$

Thus H = K, and the composition one way is the identity. Now suppose that we start with L/M/K. Let H = Gal(L/M) and let N = L^H. We already know that:

$$M \subset N,$$

and so by the Tower Law it suffices to prove that:

$$[L : N] \ge [L : M].$$

As L/K is Galois, then so is L/M. But then:

$$[L : M] = |H|.$$

As H is a set of automorphisms of L/N, we have:

$$[L : N] \ge |H| = [L : M].$$

Thus M = N and the composition the other way is the identity. Thus we have a correspondence. We have already seen that this correspondence is inclusion reversing.

The rest of the course will be adressed to deriving consequences of the Fundamental Theorem.

Theorem:

1. Primitive Element Theorem. Let L/K be a finite separable extension.

   Then L/K is primitive.

Proof: It suffices to prove that there are only finitely many intermediary fields. To this end, we are certainly free to enlarge L. Replacing L by its normal closure, we may as well assume that L/K is Galois. In this case the Galois group G is finite and so there are clearly only finitely many subgroups of G. But by the Galois correspondence there are then only finitely many intermediary fields.

It is traditional in the statement of the Fundamental Theorem to characterise when M/K is normal in terms of the associated subgroup H of G.

2. The Fundamental Theorem of Galois Theory: bis. Let L/K be a finite Galois extension. Then there is an inclusion reversing bijection between the subgroups of the Galois group Gal(L/K) and intermediary subfields L/M/K. Given a subgroup H, let $M = L^H$ and given an intermediary field L/M/K, let H = Gal(L/M).

   Furthermore M/K is normal if and only if H is normal in G. In this case the Galois group of M/K is isomorphic to G/H.

   Proof: We have already established the existence of the correspondence. Recall that M/K is normal if and only if for every $\varphi \in G$, $\varphi(M) \subset M$ if and only if $\varphi(M) = M$. Suppose that M/K is normal. Define a map:

   $$f : G \to \operatorname{Gal}(M / K),$$

   by sending φ to the restriction ψ of φ to M. As M/K is normal, ψ is indeed an automorphism of M/K. It is easy to check that f is a homomorphism of groups. Clearly H is the kernel, so that H is indeed normal. It follows that G/H is isomorphic to a subgroup of Gal(M/K). But,

   $$\begin{aligned} |\operatorname{Gal}(M / K)| &= [M : K] \\ &= |L : K]/[L : M] \\ &= |G|/|H|, \end{aligned}$$

   where we used the Tower Law and the fact that L/M and L/K are Galois extensions, so that in fact

   $$\operatorname{Gal}(M / K) \simeq G / H.$$

Now suppose that H is normal. Suppose that m ∈ M and let $\phi \in G$. Set $n = \phi(m)$. As H stabilises m, then $H = \phi H \phi^{-1}$ stabilises n. But, by the correspondence already established, the only elements of L stabilised by all of H, are in fact elements of M. Thus n ∈ M, and $\phi(M) \subset M$. But then M/K is normal.

We now turn to the problem of computing Galois groups, in explicit cases. It turns out that this is a relatively straightforward problem.

Examples:

1. Let L= $\mathbb{C} / \mathbb{R}$

Then L/K is Galois (indeed it is quadratic) and the Galois group has order two. $L = \mathbb{R}(i)$. i. i is a root of $x^2 + 1$, irreducible, and any automorphism of L/K must send i to another root. There are only two possible roots, ± i. One gives the identity, the other is complex conjugation.

Similarly, for any other quadratic extension.

2. We compute the Galois group of $x^4 - 2$ over $K = \mathbb{Q}$.

By this we mean we look at the splitting field L/K of $x^4 - 2$ and compute this Galois group. We first start by computing the splitting field. First we need to attach a fourth root of 2. Suppose that we call this α. We get M = K(α) and as $x^4 - 2$ is irreducible this is a degree four extension. Then we add a primitive fourth root of 1. Call this i, so that $i^2 = -1$. We can always pick α so that it is real. Then $i \notin M$, so that L/M is a degree two extension. Thus, by the Tower law, we have a degree eight extension. Now L/K is Galois, as $x^4 - 2$ is separable (characteristic zero).

Thus |G| = 8. We look for generators and relations. First consider the extension $L = M(i)/M$. This has order two, and so there is an automorphism τ fixing M (that is fixing α) which switches i and –i. Thus one automorphism of L/K is given by τ, where the action of τ on generators is:

$\tau\ (\alpha) = \alpha\ and\ \tau\ (i) = -i.$

Second note that there is an automorphism σ of L/K which sends α to any other root, in particular iα. Now σ need not fix i. If it does not, then τ σ fixes i and sends α to –iα. But then the cube send α to iα. Thus we may assume:

$\sigma(\alpha) = i\alpha$ and $\sigma(i) = i.$

Now $\sigma^4 = 1$ and $\tau^2 = 1$. In particular σ and τ generate two subgroups of G, of order four and two. It follows that σ and τ are generators of G. What are the relations? It suffices to compute the conjugate of σ by τ , τ σ τ . This sends i to i and α to –iα. Thus τ σ τ –1 = σ 3 . Thus G has presentation:

Generators: σ and τ.

Relations: $\sigma 4 = \tau 2 = e,\ \tau\sigma\tau - 1 = \sigma 3.$

We recognize this as D_4, the Dihedral group of order eight, the symmetries of the square. In fact we can even see the square. The action of G is determined by its action on the four roots ± α and ± iα, which are arranged in a square on the argand diagram.

Now we list all the possible subgroups and intermediate fields. We have already computed all possible subgroups of D_4. Let H be a subgroup. Then the order of H divides the order of G, so that the order of H is 1, 2, 4 or 8. The cases 1 and 8 are easy and obvious, the trivial subgroup and the whole of D_4.

Suppose that the order of H is two. Then H is generated by an element of order two. There are five of these, the two diagonal flips, the two side flips, and rotation by 180°. Thus H is one of:

$\langle\tau\rangle, \langle\sigma^2\tau\rangle, \langle\sigma\tau\rangle, \langle\sigma^3\tau\rangle, \langle\sigma^2\rangle.$

Now suppose that the order of H is four. One possibility is $\langle\sigma\rangle$. Otherwise we have to combine two elements of order two together. Now note that any subgroup of G of order 4 has index 2 and any subgroup of index two is normal. Thus if H is a subgroup of order 4 then it is normal in G. But a subgroup H of G is normal if and only if it is a union of conjugacy classes. Now the diagonal flips and the side flips are conjugates of each other, so we can combine two side flips, or two diagonal flips, but we cannot mix side and diagonal flips. Thus the subgroups of order 4 are:

$$\langle\sigma\rangle,\ \langle\tau,\sigma^2\rangle,\ \langle\sigma\tau,\sigma^2\rangle.$$

Now we list the corresponding fixed fields. Two extremes are $\mathbb{Q}$ and $\mathbb{Q}(\alpha,i)$, corresponding to G and {e}. If $M/\mathbb{Q}$ is quadratic then the corresponding subgroup H has order 4 and index 2. Thus we are looking for three subfields of degree two,

$$\mathbb{Q}(i),\ \mathbb{Q}(\sqrt{2}),\ \mathbb{Q}(i\sqrt{2}).$$

A little thought shows that in fact these fields are in the order corresponding to the subgroups of order 4.

Now let us search for the intermediate fields of order four. These correspond to subgroups of order two, so we are looking for five such fields. One obvious one is $\mathbb{Q}(\alpha)$, corresponding to τ. Similarly $\mathbb{Q}(i\alpha)$ corresponds to $\sigma^2\tau$. Also there is $\mathbb{Q}(i,\sqrt{2})$ corresponding to σ^2. Now we need to compute the intermediate field associated to στ. There is not much to do but write down the general element of L and see when it is fixed by στ. Note that we know M contains $\mathbb{Q}(i\sqrt{2})$, so we only need to write down a basis for $\mathbb{Q}(\alpha, i)/\mathbb{Q}(i\sqrt{2})$, which is easily seen to be:

$$1,\ \alpha,\ i,\ i\alpha.$$

So the general element of $\mathbb{Q}(\alpha, i)$ is:

$$a + b\alpha + ci + d(i\alpha)$$

and this is sent to:

$$a + bi\alpha - ci + d(\alpha).$$

So c = 0 and b = d. Thus the corresponding fixed field is $\mathbb{Q}((1+i)\alpha)$. Similarly, the other field of degree four is $\mathbb{Q}((1-i)\alpha)$.

Now we turn to the problem of computing Galois groups over finite fields. It turns out that this problem is almost completely trivial.

Let R be a ring of characteristic p. The map:

$$\Phi : R \to R$$

defined as:

$$\Phi(a) = a^p,$$

is a ring homomorphism, called the Frobenius map.

Theorem: Let L/K be an extension of finite fields.

Then L/K is Galois. Moreover the Galois group is cyclic, generated by a power of Frobenius.

Proof: We already know that $L \simeq \mathbb{F}_q$, $K \simeq \mathbb{F}_r$, where q and r are powers of p, $q = p^n$ and $r = p^m$, where m|n and d = n/m is the degree of the extension. Moreover L is the splitting field of the polynomial $x^q - x$ and the non-zero elements of L are precisely the roots of unity. As Φ is injective and L is finite, Φ is clearly an automorphism of L. The fixed field of Φ^k is the set of all elements of L such that:

$$a^t = a,$$

where $t = p^k$, that is, all roots of the polynomial :

$$x^t - x.$$

But then Φ^m fixes K and the smallest power of Φ^m that fixes L is d. Thus,

$$\langle \Phi^m \rangle$$

is a subgroup of the Galois group of order d. But the Galois group has order d.

Example: Let us compute the Galois group of $f(x) = x^4 + x + 1$ over the field $\mathbb{F}_2$.

The problem essentially boils down to factoring f. f certainly does not have any linear factors as it has no roots. Suppose then that f were reducible. Then,

$$x^4 + x + 1 = (x^2 + ax + b)(x^2 + cx + d),$$

where a, b, c and d $\in \mathbb{F}_2$. Looking at the coefficient of x^3, we have c = −a = a and looking at the constant coefficient we have bd = 1, so that b = d = 1. Thus we would have,

$$x^4 + x + 1 = (x^2 + ax + 1)(x^2 + ax + 1) = x^4 + ax^2 + 1,$$

a contradiction.

Thus $x^4 + x + 1$ is irreducible. Let L be a splitting field and let α be a root of f. Then

$$M = \mathbb{F}_2(\alpha)$$

is normal, as all extensions of finite fields are normal. Thus L = M and L/F2 has degree four. Thus the Galois group is cyclic of order 4.

Lemma: Let $f(x) \in \mathbb{R}[x]$ have odd degree.

Then $f(x)$ has a real root.

Proof: We may as well suppose that $f(x)$ is monic. We may write,

$$f(x) = x^n + \sum a_i x^i,$$

Let $m = \max |a_i|$, and pick $x > nm$. Then,

$$\begin{aligned} f(x) &\geq x^n - \left|\sum a_i x^i\right| \\ &\geq x^n - \sum |a_i||x^i| \\ &\geq x^n - nmx^{n-1} \\ &\geq x^{n-1}(x - nm) > 0. \end{aligned}$$

Similarly, $f(x) < 0$, for $x < -nm$. Thus $f(x)$ must have a zero by the Intermediate Value Theorem.

Theorem: (Fundamental Theorem of Algebra). $\mathbb{C}$ is the algebraic closure of $\mathbb{R}$.

In particular $\mathbb{C}$ is algebraically closed.

Proof: Let $L/\mathbb{C}$ be a finite extension. It suffices to prove that $L = \mathbb{C}$. Passing to a normal closure we may assume that $L/\mathbb{R}$ is Galois. Let G be the Galois group. Let H be a Sylow 2-subgroup. Let $M = L^H$ be the corresponding fixed field.

Then $M/\mathbb{R}$ has odd degree. Let $\alpha \in M$ and let $f(x)$ be the minimum polynomial of α. Then $f(x)$ has odd degree. But then $f(x)$ has a root in $\mathbb{R}$. As $f(x)$ is irreducible, its degree must be one. But then $\alpha \in \mathbb{R}$ and so $M = \mathbb{R}$. Thus $L/\mathbb{R}$ has degree a power of two. Similarly for $L/\mathbb{C}$.

Suppose that G is not the trivial group. By Sylow's Theorem there is a subgroup H of G of index two. But then there would be an extension $M/\mathbb{C}$ of degree two. As the characteristic is zero, there would then be an element $\alpha \in M$ such that $\alpha^2 \in \mathbb{C}$. But C is certainly closed under taking square roots, a contradiction.

Thus G is trivial and $L = \mathbb{C}$.

The Hilbert Basis Theorem

A ring R is said to be a Noetherian ring if it satisfies the ascending chain condition, that is, for all ascending chains of ideals $I_1 \subseteq I_2 \subseteq \ldots \subseteq I_n \subseteq \ldots$ there exists an $N \in \mathbb{N}$ such that for all $m \geq N$ we have that $I_m = I_N$. Equivalently, we proved that R is Noetherian if and only if every ideal I is finitely generated, that is, there exists $x_1, x_2, \ldots, x_n \in I$ such that:

$$I = (x_1, x_2, \ldots, x_n).$$

We about to prove a very important result known as the Hilbert basis theorem which tells us that if R is a Noetherian ring then the corresponding ring of polynomials of a single variable $x, R[x]$. is a Noetherian ring. We first need the following lemma.

We first need to get some notation out of the way. If $F \in R[x]$ then F is of the form:

$$F(x) = a_n x^n + a_{n-1} x^{n-1} + \ldots + a_1 x + a_0$$

We define the function $\text{cof}(F)$ to be the leading coefficient of F. That is, if F has degree n as above then:

$$\text{cof}(F) = a_n$$

For ≥ 0 and for an ideal I. we define:

$$J_m = \{\text{cof}(F) : F \in I, \deg F \leq m\}$$

Lemma: Let R be a Noetherian ring and let I be an ideal. Then for all $m \geq 0, J_m$ is an ideal.

- Proof: Let $a \in J_m$ and let $b \in R$. Then $a = \text{cof}(F)$ for some polynomial $F \in I$ with $\deg F \leq m$. Consider the function bF. Then $bF \in I$ since I is an ideal. Furthermore, $\deg F \leq m$. Therefore $\text{cof}(bF) \in J_m$. But $\text{cof}(bF) = ab$. So, $ab \in J_m$.
- Now let $a, b \in J_m$. Then $a = \text{cof}(F)$ and $b = \text{cof}(F)$ for some polynomials $F, G \in I$ with $\deg G \leq m$ Let $\deg F = s$ and let:

$$F(x) = ax^s + a_{s-1}x^{s-1} + \ldots + a_1x + a_0$$
$$G(x) = bx^t + b_{t-1}x^{t-1} + \ldots + b_1x + b_0$$

- Without loss of generality, assume that $s \geq 1$ Define a new polynomial H by:

$$H(x) = F + x^{s-t}G)$$

- Since $F, G \in I$ we have that $H \in I$. Furthermore observe that:

$$\begin{aligned} H(x) &= [ax^s + a_{s-1}x^{s-1} + \ldots + a_1x + a_0] + x^{s-t}[bx^t + b_{t-1}x^{t-1} + \ldots + b_1x + b_0] \\ &= ax^s + a_{s-1}x^{s-1} + \ldots + a_1x + a_0 + bx^s + b_{t-1}x^{s-1} + \ldots + b_1x^{s-t+1} + b_0x^{s-t} \end{aligned}$$

- Therefore $\text{cof}(F + x^{s-t}G) = a + b$ Hence $(a+b) \in J_m$ Thus J_m is an ideal.

Theorem: (The Hilbert Basis Theorem): Let R be a Noetherian ring. Then $R[x]$ is a Noetherian ring.

- Proof: Let $I \subseteq R[x]$ be an ideal and for each $m \geq 0$ let J_m be defined in terms of I.
- Consider the following ascending chain of ideals:

$$J_1 \subseteq J_2 \subseteq \ldots \subseteq J_n \subseteq \ldots$$

- Since R is Noetherian, there exists an $N \in \mathbb{N}$ such that for all $m \geq N$ we have that $J_m = J_N$.
- Also, since R is a Noetherian ring every ideal in R is finitely generated. So for each $m \geq 0$ there exists elements $a_{m_1}, a_{m_2}, \ldots, a_{m_k}$ such that:

$$J_m = (a_{m_1}, a_{m_2}, \ldots, a_{m_k})$$

For each $1 \le j \le k$, choose a polynomial $F_{m_j} \in I$ such that $a_{m_j} = \mathrm{cof}(F_{m_j})$. Let I' be defined as the ideal generated by the $F_{m_j}s$ for all $m \le N$. We claim that I I'.

- By definition, we have that $I' \subseteq I$. Now suppose that $I' \not\supseteq I$. Then there exists a function $G \in I$ such that $G \notin I'$ Let G be chosen of minimal degree in I' and let $\deg(G) = d$ Then $\mathrm{cof}(G) \in J_d$ So $\mathrm{cof}(G)$ has the form:

  $$\mathrm{cof}(G) = \sum \alpha_j \,\mathrm{cof}(F_{d_j})$$

  Where $a_j \in R$ and F_{d_j} are in the list of generators for I' and where $\deg(F_{d_j}) \le d$.

  $$Q(x) = \sum \alpha_j x^{d-\deg(F_{d_j})} F_{d_j}$$

- Then $Q \in I$ and $\deg(Q) = d$. But $\mathrm{cof}(Q) = \mathrm{cof}(G)$ .Then $G - Q \in I$. But $\deg(G-Q) \le d-1$. Since $G \notin I'$ and $Q \in I'$ we hae that $G - Q \notin I'$ and is such that $\deg(G-Q) < d$ But this contradicts G having minimal degree in I.

- Therefore $I = I'$. Since is finitely generated so is I. So every ideal in $R[x]$ is finite generated, i.e., $R[x]$ is Noetherian.

The Isomorphism Extension Theorem

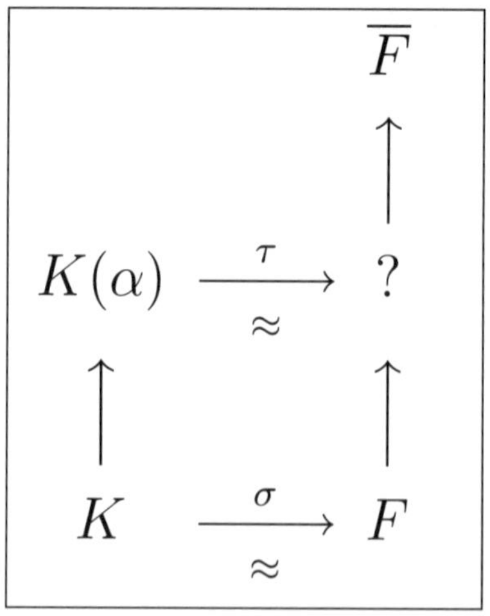

Let K(α) be a simple algebraic field extension of K, let σ : K → F be an isomorphism of K onto a field F with algebraic closure $\overline{F}$. Let p(x) = $\mathrm{mipo}_{\alpha,K}$ (x) and consider the irreducible and monic polynomial $q(x) = \overline{\sigma}(p(x)) \in F[x]$, where $\overline{\sigma} : K[x] \to F[x]$ is the isomorphism sending $x \mapsto x$ induced by σ : K → F.

Then every isomorphism τ mapping K(α) onto a subfield of F and extending σ, maps α onto a root β of q(x) and maps K(α) onto F(β). Moreover, for every root β of q(x) there exists exactly one isomorphism τ of K(α) onto a subfield of F which extends σ and maps α onto β.

This was our construction:

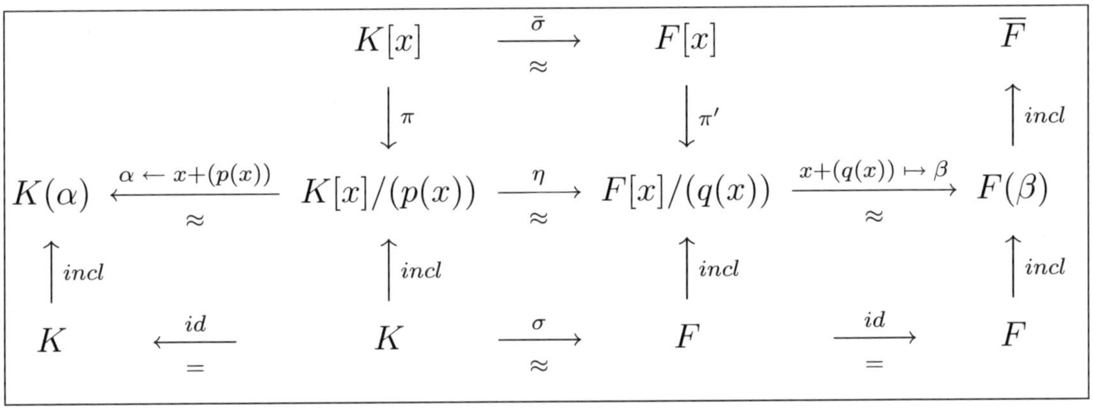

We now extend this result to general algebraic extensions using Zorn's Lemma:

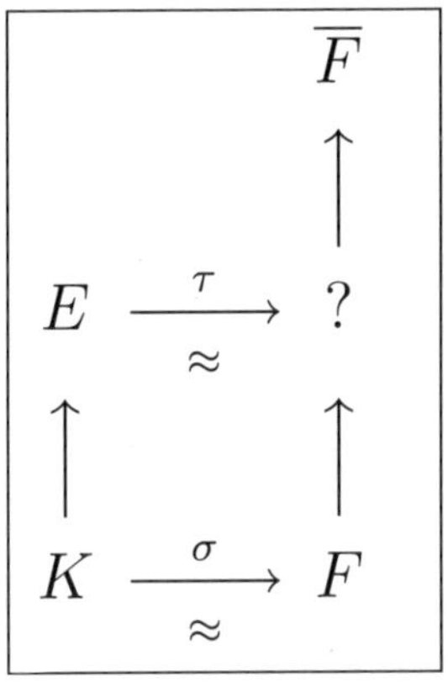

Theorem: Let E be any algebraic field extension of K and let $\sigma : K \rightarrow F$ be an isomorphism of K onto a field F with algebraic closure $\overline{F}$. Then σ extends to an isomorphism τ of E onto a subfield of $\overline{F}$.

Proof: Consider the collection $\mathscr{L}$ of all pairs (L, λ) where $\mathscr{L}$ is a field such that $K \subseteq L \subseteq E$ and **λ** is an isomorphism of L onto a subfield of $\overline{F}$ which extends σ. Then $\mathscr{L}$ is not empty, because it contains (K, σ). We can partially order $\mathscr{L}$ by declaring $(L_1, \lambda_1) \prec$ (L2, λ2) if and only if $L_1 \subseteq L_2$ and λ_2 extends λ_1. Then every chain C in $\mathscr{L}$ has an upper bound $(\tilde{L},\tilde{\lambda})$ in $\mathscr{L}$. Indeed, as we have seen in similar proofs before, if $\mathscr{C}$ is a nonempty chain in $\mathscr{L}$, we can take for $\tilde{L}$ the union of all fields L with $(L, \lambda) \in \mathscr{C}$ and we may define $\tilde{\lambda}$ (a) = λ(a) when $a \in L$ with $(L, \lambda) \in C$. Therefore, by Zorn's Lemma, $\mathscr{L}$ contains a maximal element (M, τ). The proof will be complete once we show that M = E.

Suppose then, to the contrary, that $K \subseteq M \subseteq E$. Then there is an element $\alpha \in E$ with $\alpha \notin M$. Consider $M \subseteq M(\alpha) \subseteq E$ and put N = τ [M]. Then $F \subseteq N \subseteq \overline{F}$. Since F is an algebraic extension of N and since $\overline{F}$ is algebraically closed, we have $\overline{N} = \overline{F}$. Since $\tau : M \rightarrow N$ is an isomorphism of M onto N and since M(α) is a simple algebraic extension of M, there is an isomorphism ξ from M(α) onto a subfield of $\overline{N} = \overline{F}$ which extends τ. (This follows from the special case of this theorem proved in class for simple extensions.) Since τ extends σ, so does ξ. Therefore, $(M(\alpha), \xi) \in \mathscr{L}$ with $(M, \tau) \prec (M(\alpha), \xi)$ and $(M, \tau) \neq (M(\alpha), \xi)$. But this contradicts the maximality of (M, τ).

References

- Node7, NM, lectures: eng.hmc.edu, Retrieved 14 July, 2019
- Cayley-Hamilton-theorem: encyclopediaofmath.org, Retrieved 11 January, 2019
- The-fundamental-theorem-of-ring-homomorphisms: mathonline.wikidot.com, Retrieved 29 June, 2019
- The-hilbert-basis-theorem: mathonline.wikidot.com, Retrieved 30 April, 2019

Permissions

All chapters in this book are published with permission under the Creative Commons Attribution Share Alike License or equivalent. Every chapter published in this book has been scrutinized by our experts. Their significance has been extensively debated. The topics covered herein carry significant information for a comprehensive understanding. They may even be implemented as practical applications or may be referred to as a beginning point for further studies.

We would like to thank the editorial team for lending their expertise to make the book truly unique. They have played a crucial role in the development of this book. Without their invaluable contributions this book wouldn't have been possible. They have made vital efforts to compile up to date information on the varied aspects of this subject to make this book a valuable addition to the collection of many professionals and students.

This book was conceptualized with the vision of imparting up-to-date and integrated information in this field. To ensure the same, a matchless editorial board was set up. Every individual on the board went through rigorous rounds of assessment to prove their worth. After which they invested a large part of their time researching and compiling the most relevant data for our readers.

The editorial board has been involved in producing this book since its inception. They have spent rigorous hours researching and exploring the diverse topics which have resulted in the successful publishing of this book. They have passed on their knowledge of decades through this book. To expedite this challenging task, the publisher supported the team at every step. A small team of assistant editors was also appointed to further simplify the editing procedure and attain best results for the readers.

Apart from the editorial board, the designing team has also invested a significant amount of their time in understanding the subject and creating the most relevant covers. They scrutinized every image to scout for the most suitable representation of the subject and create an appropriate cover for the book.

The publishing team has been an ardent support to the editorial, designing and production team. Their endless efforts to recruit the best for this project, has resulted in the accomplishment of this book. They are a veteran in the field of academics and their pool of knowledge is as vast as their experience in printing. Their expertise and guidance has proved useful at every step. Their uncompromising quality standards have made this book an exceptional effort. Their encouragement from time to time has been an inspiration for everyone.

The publisher and the editorial board hope that this book will prove to be a valuable piece of knowledge for students, practitioners and scholars across the globe.

Index